Railway Modelling In Profile :

WHERE TO P
YOUR
LAYOUT

by

NIGEL ADAMS

"Wenfordbridge Goods Depot". A view of John Greenwood's beautifully built 2mm finescale layout. ***John Greenwood***

ISBN 1 900298 24 4

Layout, Design and Typeset by Peter Midwinter
The Railway Study Centre

Printed by GWASG DWYFOR, Pen-y-groes,
Caernarfon, Gwynedd, LL54 6DB.

Published by Cheona Publications
The Railway Study Centre
Tal Eithin Isaf
Llanllyfni
Caernarfon
Gwynedd LL54 6RT

Tel: 01286 882 332 Fax: 01286 882 062

"Stoke Fleming". A view of Alan Searle's layout with the Auto-train about to depart for Kingsbridge. ***Alan Searle***

Justin Paul Engineering Ltd. on the author's Bottrill Street Yard layout. The idea for the name of the factory was cribbed from Revd. Peter Denny who named a factory on his 'Buckingham Branch'"Crispin Paul Bros" after his two sons. *Martin Hewitt*

Foreword.

Getting started in railway modelling today can be quite intimidating, for the raw beginner (young or old) especially. One can argue that 'we've never had it so good' (particularly in 'OO' gauge) with regard to what the main RTR manufacturers offer us, at quite fantastic value prices, 'off the shelf'. However, despite this sea of riches, what the newcomer sees in magazines or at exhibitions, where the work of the greatest current exponents in the hobby is on display, can be as frightening as it is inspirational. Not just the absolute starter either, for the hobby is awash with those returning to it, now, maybe, in middle-age, with children fled the nest, some spare time, a bit of spare cash and some little space at last. Thus then, we have a bit of a paradox in the current state of the hobby, where what's available to the modeller has never been better, but perhaps in parallel with (or even because of) this, the standards seem so high as to be unobtainable for most mere mortals.

This book will be a great help in this respect. The author has a long (and most pragmatic) pedigree in the hobby and has built numerous exhibition layouts in the simplest, most sensible manner. It is presented in a typical 'no nonsense' manner, full of useful ideas and examples of how to make layouts, where to put them, how to present them and how to operate them. All examples, from many different sources and builders, are practical, achievable solutions of how to make a successful model railway. Though it's aimed at those making a small layout, this book is also most useful for those contemplating expanding their systems.

In short, I am delighted to commend this volume to anybody involved in this most wonderful hobby (there is none better), and I am privileged to supply this foreword and also some of the photographs for it. I have worked with Nigel for quite some years now, photographing his layouts and being involved with much of his published work. Read this book, take it in, use it (whatever your experience) and become involved in, more involved in or just return to this most satisfying and creative of all hobbies. This volume will give you a great start.

Tony Wright
Photographer & Assistant Editor.
British Railway Modelling

Above. "The Shed Mk 1". A view showing how the entrance from the fiddle-yard was disguised. ***Martin Hewitt***

Below. The low relief backscene on the author's 009 **"Small M.P.D."** showing how effectively the Airfix (now Dapol) Engine Shed Kit can be used. The bottom three courses of bricks were removed to allow for the fact that the rolling stock was that much smaller than standard gauge stock. ***Author's collection***

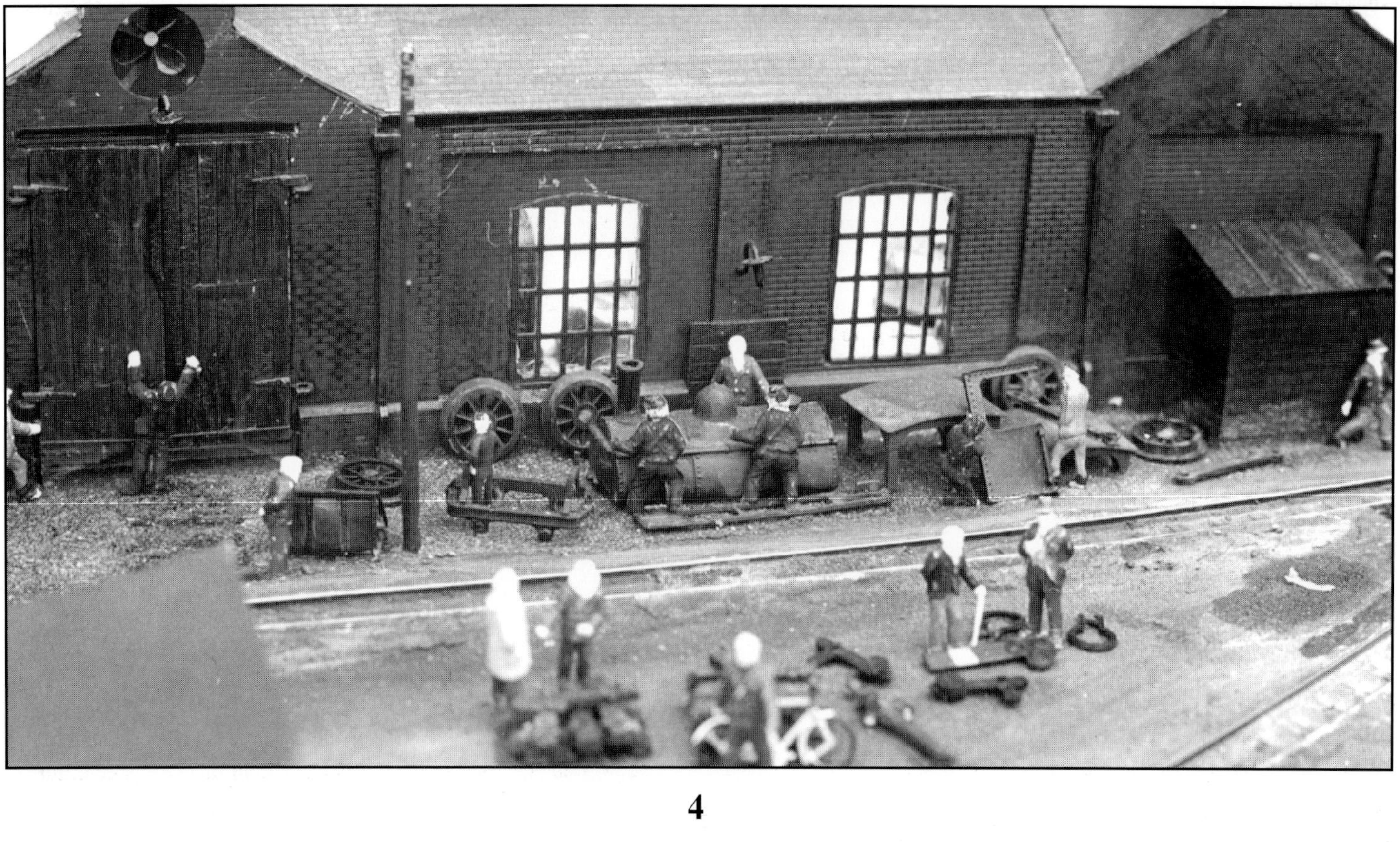

Amberwoods. A photograph of a small 009 layout built by the author in the mid 1970's. ***Brian Monaghan***

WHERE TO PUT YOUR LAYOUT

Introduction.

So you want to build a layout!

The first question is where you are going to put it and that obviously depends on the space available in your home.

Until you have decided the best place to locate your layout you cannot really decide on other important questions such as the scale and the gauge and whether it is going to be fixed or portable.

I model in 'O' gauge and people often think that 'O' gauge needs a large area but that is not so. I have built two very satisfying layouts in a space of 3' 9" x 1' 10" (see **Photographs 1/1 & 1/2**). People also often think that they would like a large layout and then say that they haven't got the room for it.

If they genuinely want a large layout and live in a small terraced house or a flat, that may be so but, in many aspects of our life, we make compromises and use what we have available in the best way we can.

Take cars for example.

Before I was ordained I worked in the car industry and the last company I worked for was Jaguar. They produce some superb cars and, if I could afford one, I would do so. However, I cannot afford one and, in any case, I want an estate car because, amongst other things, I want to transport my layout to exhibitions. So, although I would love to own a Jaguar, I recognise that it is not possible and, instead, I currently drive a Skoda Felicia Estate Car!

It is far better to have a layout than no layout at all. You can always progress to a larger layout in the future if circumstances permit and, in the meantime, you will have gained valuable experience and built up your rolling stock. That may be when your children have grown up and a spare room becomes available or perhaps you buy a bigger house. Whatever the reason your circumstances will have changed and what was impossible becomes possible.

In any case, having a large space available can create problems of its own.

For example, my last 2 vicarages were of a relatively modern design – one built in 1977 and one in 1989. The Church

Photo. 1/1 "Small M.P.D." *Martin Hewitt*

Photo. 1/2 "The Shed" *Author*

Commissioners and the dioceses have made it policy to replace the old rambling vicarages to save on maintenance and running costs, and rightly so in my opinion. In my last vicarage my layout was in the smallest bedroom which was given over entirely to my hobby. It measured 10' 9" x 7' 10" and satisfied my needs.

One of my colleagues lived in a vicarage built in the early 1900's. It had 7 bedrooms compared with my 4 and 2 of those bedrooms were in the loft and were huge.

To fill one of those bedrooms with a layout of whatever scale would be very expensive indeed. Furthermore a large layout requires a lot of locomotives and rolling stock. This adds significantly to the cost. Perhaps most important of all a large layout takes a long time to build. This is a particular drawback if you have to move frequently for your job. There is a danger of progress being so slow, in these days of people working long hours and dealing with family commitments, that the builder of a large layout loses interest because there might be very little to see for the effort put in.

So having decided where the layout is to go, we move to the other relevant questions – scale, gauge, fixed or portable.

Often a small space means that the builder chooses to build a layout in N gauge (2mm scale) or if narrow gauge modelling is their interest, perhaps in 009. Both these scales have track which is approximately 9mm between the rails.

Probably still the most common gauge and scale is 'OO' which is 4mm to the foot and there are many different types of layout using that scale/gauge combination.

The next popular scale in size is 7mm / ft, commonly called 'O' gauge or O - 16.5 which is 7mm / ft Narrow Gauge.

There are lots of layouts in the model railway press in these scales and some to be seen at exhibitions up and down the country.

The space available will also determine whether the layout will be permanently erected or whether it will be stored in some way. We will look at those options in more detail later.

You also need to decide if you want to exhibit your layout. Speaking as someone who has been exhibiting layouts for over 35 years, and attending shows for much longer, that is a must for me.

Therefore, even if I had a large room available, I would not want to fill it with a layout because it would be almost impossible to take such a layout to a show.

Generally, large exhibition layouts are built and shown by a group of people who may well be in a club. Such layouts take a lot of time to build and have to be dismantled, transported to the exhibitions and re-erected. All this takes time and money because there is a need to hire a large van unless one of the group owns one or knows someone who does.

From an exhibition manager's viewpoint, the large layout is usually expensive due to the transport costs and the number of operators required to run it throughout the show.

So, look around your home, decide where the layout can go, whether it is to be erected permanently or stored, whether it is to be exhibited, what scale and gauge you wish to use and then begin to think about the planning and design aspects.

Finally, let me say that we are all different, I like building layouts and prefer them to be relatively small because I wish to exhibit them and transport them in my own car.

Other people may prefer to build a large layout over a period of time and never exhibit it. For those people, a large room is a god-send and they work to a long term plan, producing a marvellous layout. But, whatever sort of person you are, you cannot build a large layout unless you have the space and that is the controlling factor.

2. What is home?

As the first requirement for a layout is space, perhaps we should briefly look at the various types of home people live in.

Young modellers living at home have no choice but to negotiate with their parents as to what, if any, space can be allocated to their model railway hobby. Very often it is a relatively small space in their bedroom and that may be just as well because, in some ways, space equates to money.

Generally it does not cost as much to build a small layout as it does a large one and young people who are not working may not have much money to spend on their hobby.

The same principle applies when we join the housing market.

Because house buying is expensive most people start off in a small house or a flat and "trade up" when they can afford it. The same principle applies to buying a car.

So both house builders and car manufacturers have designs which appeal to the "bottom end of the market" which is where people usually start.

So a home can be a mobile home, a flat, a maisonette, a small terraced house, a large terraced house, a semi-detached house (large or small), a detached house (large or small) or even a canal boat. A home may also be a bungalow or a 2 or 3 storey house. It may or may not have a garage. If it has, that garage may be an integral part of the house, it may be at the bottom of the garden, it may be alongside the house or it may be in a block of garages.

A home may or may not have a garden. If it has it may or may not have room for a garden shed.

A home may be an old house with large rooms and a loft or a basement or it may be a small modern house with a loft which has those W shaped roof trusses which often make the loft virtually unusable for our purposes.

We also buy different sorts of houses in our life to suit our work and personal circumstances.

We may start off in a small "starter" house. These can be for single people or couples. As children come along and we perhaps get pay increases or promotion at work, we move to a bigger house.

Unless job changes dictate otherwise, we may stop in the larger semi-detached or detached house until the children have grown up and left home and then retirement begins to loom on the horizon – maybe even early retirement or, worse still, redundancy – and we begin to think we no longer need such a large house and so we move into a smaller house or a bungalow for the last years of our life. So, over the years, we may live in a number of different types of house and the type of house we live in has a huge bearing on the space available for our hobby.

Also, it seems that people have to move jobs more often these days and to move a large permanent layout is a nightmare. The chances of finding another house with a room which will take such a layout – even if it has been designed with moves in mind – are not very good.

My own experience in 39 years of married life has seen us live in the following types of property:

- 2 bedroom upstairs maisonette (newly built in 1964)
- 3 bedroom semi-detached (newly built in 1966)
- large 3 bedroom semi-detached (built in 1939)
- 4 bedroom detached (newly built in 1974)
- 4 bedroom detached (rented at Theological College)
- 3 bedroom semi-detached curate's house (extended pre-war house)
- 2 modern 4 bedroom vicarages (of very different designs)

and we now have a 3 bedroom end of terrace cottage in which we live in retirement (almost next to the Talyllyn Railway in Tywyn!).

That last move was quite traumatic in many ways because we had to drastically prune our furniture, books, pictures etc. as well as my model railway possessions. Just to make storage more of a problem it does not have a garage. So the house in which we live has a huge bearing on what sort of layout we can build.

3. Where can the layout go?

The first question which has to be decided is indoors or outdoors. If you live in a block of flats and do not have a garden, that question is quickly answered. The layout has to be indoors.

If you have a garden you may be able to have a layout round the garden or a layout in a shed. If you have a garage (free standing or integral) a layout could be located there. In fact there are many people, security implications apart, who would argue that, for all sorts of reasons, your car should not be kept in the garage. If you are one of those people then your garage offers a large potential space for a layout. There are disadvantages too and we will consider garages and sheds in Chapter 5.

Having looked very briefly at outdoor possibilities we turn now to the indoor ones.

Utopia is obviously to have your own Railway Room. If this is possible you can work on the layout or operate it without wasting time setting it up. Obviously the big disadvantage of all portable layouts is that you need to remove all the stock and put away the baseboards. So to work on it you have to get the boards out again, unless of course, you just want to work on one section.

The "dedicated" room enables you to operate or work on the layout for a time, and then leave it exactly as you wish and shut the door. This is particularly true if you operate the layout to a timetable. You can go in and start where you left off. Another advantage of a railway room is that you can easily store stock, tools, unmade kits etc.

However, realistically, most modellers are unlikely to find Utopia straight away! So what are the alternatives?

Let's have a brief look at the various possibilities and I will expand on these in Chapter 4.

Having said that the sole use of a spare room is Utopia, let's start there on the basis that the spare room is precisely what it says - "spare".

Fig. 3/1 Plan of my 'modelling room'

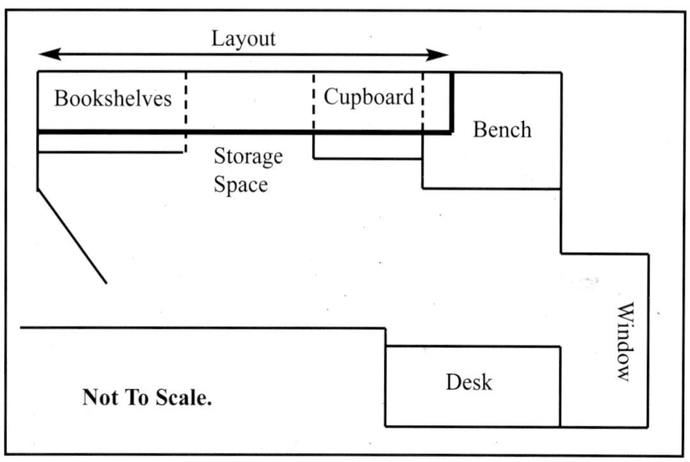

Photo. 3/1
A view of Rye Town. Look at the wealth of fine detail that is included in this portrayal of a small country station.
Tony Wright Courtesy B.R.M.

As I said earlier, we retired to an end of terrace 3 bedroom cottage. The smallest of the bedrooms is the spare room and it is L-shaped **(See Figure 3/1)**. In that room, with the full agreement of my wife, I have located my layout, workbench and computer work-station and use the space underneath the layout for storage. There is still room (just!) to put a folding camp-bed in there if necessary. So far it has not been needed but the facility is there.

Sticking with bedrooms, there are various other possibilities. One is to have a permanent layout and use the space underneath for storage. A fellow member of the Oxford M.R.C. in the 1960's did this and fitted in the necessary furniture. He could even sit in bed and operate the layout!

The second bedroom in our cottage is ideally suited to have a layout in it with storage underneath.

Before we had built-in wardrobes fitted in the largest bedroom, there was an ideal alcove for a small layout and it would still be possible to have a small narrow layout on one wall (if the storage heater was relocated). There have been various examples of layouts in bedrooms described in the model railway press and a few are listed below:

Foldaway layout – Railway Modeller November 1981.
Fold and store under the bed (Rye Town) – British Railway Modelling June 1994. **(Photo 3/1)**
Store under the bed (Waterloo) – Railway Modeller November 1985
Bedroom Branchline – Railway Modeller April 1980
Mini bedroom layout – Railway Modeller June 1985
Layout under a Captain's Bunk – Railway Modeller January 1990.
A railway for Neil's Bedroom – Railway Modeller February 1986.

Remaining upstairs in our cottage, it would be perfectly possible to have a narrow layout (say 9" - 10" wide) on the landing with a removable fiddleyard if necessary. The space underneath the layout could be used for storage, especially if it was about 45" - 48" high. One could use bookshelf kits sold by stores like Focus, B&Q etc. as the basis for the layout if the landing was wide enough. Alternatively, you could use these kits for any long and thin layout.

The loft or the attic is another favourite place for a layout and, depending on the type of roof supports and the design of the loft, you can get a large layout in.

Obviously, the loft is susceptible to extremes of heat and cold and needs to be properly insulated and ventilated and to have a power supply and heating.

A word of warning. If you are converting your loft you need to have expert advice – even if only whether the beams will support the floor and people walking on it – and you may need planning or building regulation approval. Check with a builder or your local authority to be sure.

Older houses tend to have larger supporting timbers and plenty of headroom. Certainly my elder son's mid 1930's house has an ideal loft area and my younger son has had his loft professionally converted into a bedroom.

Lighting and access are also important factors. Good access means that the loft area can also be used for storing other household items which are not regularly used (e.g. suitcases).

The design of the roof is key to the use of the loft. Our 1974 built house had mass – produced timber W frames. It is still possible to have a layout in the loft if the measurements of the W allow it. We built one for my sons in ours but it is not ideal by any means. In my last 2 vicarages – both relatively modern – the pitch of the roof was high and enabled my elder son to have both his layouts and his work bench in the loft as well as storing all the other things people store in lofts.

The noise level downstairs is improved by laying (old?) carpet down on the floor as it deadens the sound. There are articles in the railway modelling magazines from time to time describing "loft layouts" and how to convert a loft.

Moving downstairs now, there are again many possibilities. The design of your house may make it possible to have a long thin layout in the hall. It could be possible to have an L-shaped layout in the hall too. It obviously depends on the way the house is designed. Clearly here, as elsewhere downstairs, the finish of the layout needs to be good. Most spouses will not take kindly to a mess in the very visible part of the house.

Storage under a layout in the hall is a good idea too. You can cover the front of the layout with curtains or, alternatively,

Photo. 3/2 ***Tony Wright Courtesy B.R.M.***
A railcar and coach leaves The Wharf platform. The railcar is a Peco coach body mounted on an 'HO' Lima diesel chassis with only the front part of the bonnet left of the original locomotive.

you can build it on cupboard units. A work bench in the hall is not a good idea from a tidiness point of view. However it could be on castors and be stored under the layout and pulled out when required. Another alternative is to have the work bench in a bureau. The drop down front is the work area. That would be presentable in any room.

Turning now to another part of the hall, have you considered under the stairs?

Many years ago Cyril Freezer built a layout in that location called "Tregunna" which had a removable storage yard which came out into the hall when in use.

I seriously considered using an under stairs cupboard when I was a vicar in Coventry. My study had such a cupboard and it would have been perfectly possible to build a layout there.

In our first house I built a layout along one wall of the dining room. The storage yard rested on the table when in use. Otherwise it was kept on a shelf under the layout. In fact my wife was very happy to have the layout there because of the extra storage it gave. She made curtains to go along the front of the layout.

In the lounge there are various possibilities which could also apply in the dining room if you prefer. For example anywhere where there is an alcove has possibilities. You can build a "wall hugger" which rests on cupboards or maybe you wish to build your own units on which the layout is to rest. It is also possible to have a folding lid **(See Figure 3/2)** so that the layout is free of dust.

Layouts have also been built in TV cabinets and on coffee tables. Some people have built small layouts in glass fronted cases which can either be free standing on a piece of furniture or fixed to the wall on brackets.

Gordon Gravett built such a layout which he exhibited for a number of years. It was a 7mm narrow gauge model of the winding house at Abergynolwyn on the Talyllyn Railway.

Other possibilities anywhere in the house exist if all you want to do is to store a portable layout.

I have done this a number of times. The first time was in a 1930's semi-detached house which had large rooms.

In the dining room we had a storage box which my father-in-law made for me. The small layout slid in and out on runners and the stock boxes etc. were stored underneath. **(See Figure 3/3)** It was made of plywood and covered in Fablon. Our elder son was then only 3 and the top of the box was an ideal height for him to play on.

I also built a layout on an ironing board in the late 1960's. This had a cover like an upturned box over it and when the legs were folded it could be stored anywhere.

In my vicarage in Coventry 2 bedrooms had large built in cupboards. My elder son and I stored 2 layouts (suitably protected) there on end and simply got them out when we wanted to use them. The stock boxes were stored there too. See **Figure 3/4** for a track plan and **Photographs 3/2, 3/3, 3/4 & 3/5.**
One of the most ingenious "living space layouts" to be seen on the exhibition circuit is a 7mm scale layout built by Ian Hopkins called "St. George's Hill". To quote the heading of Model Railway Journal No. 100 1998, in which it was described "Ian Hopkins' model of suburban Surrey is far more than a 'funny little layout in a grandfather clock' – it's a brilliant exercise in squeezing 7mm into a small space. The fact that it packs away into a slim, domestically acceptable cabinet has not compromised modelling quality".

It was designed to fit into a small studio flat. The area for the layout when erected at home was only 9 feet, and the boards consist of two 4ft long and one 7ins. long and the width is 14ins. If you haven't seen it you really should make the effort. If you cannot get to an exhibition where "St. George's Hill" is being shown, try and obtain a copy of MRJ No. 100. It will be well worth it, in my opinion.

Our American friends have come up with some good ideas too. Model Railroader magazine publishes two books which describe some excellent ideas.

They are entitled:
Small Railroads You Can Build. (edited by Kent Johnson)
Model Railroading in Small Spaces. (Matt Chibbaro)
They are available in the U.K.

Fig. 3/2

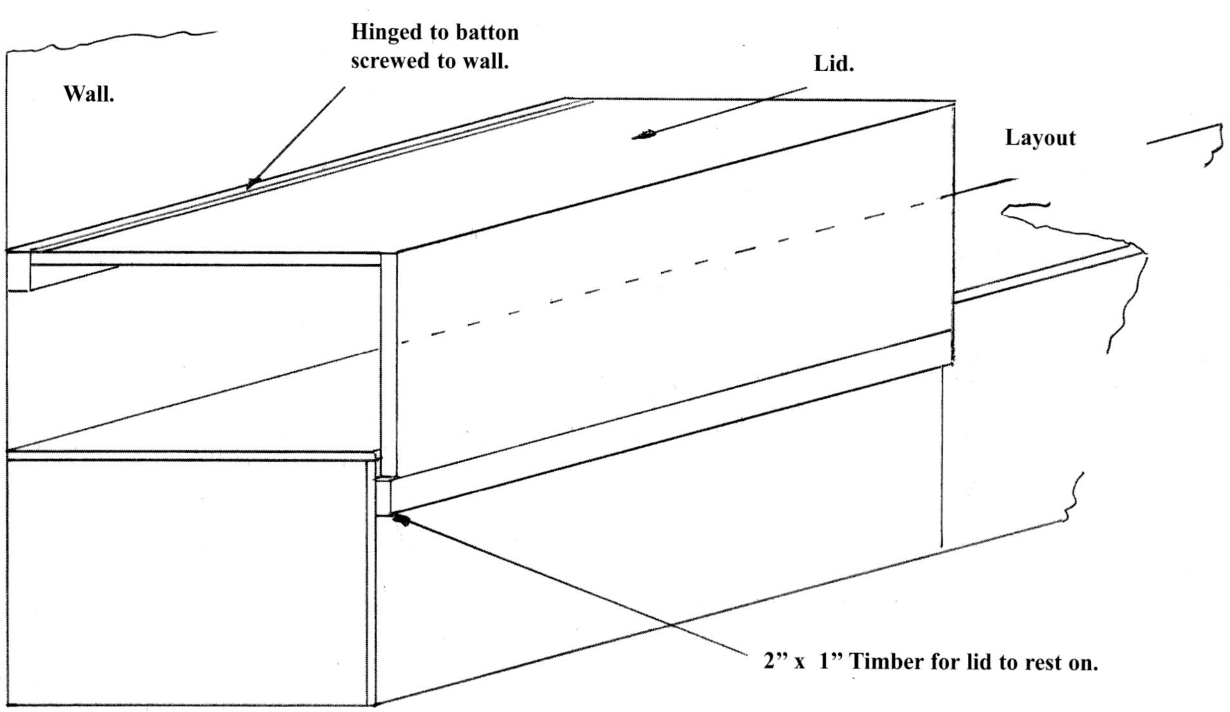

Finally in this chapter, having looked briefly inside and outside the house, and upstairs and downstairs, if all else fails you can build an extension! In today's property market that may be worthwhile financially as it should increase the value of your house. If that's too much to consider and you have a largish garden and good access, you can always consider buying a second-hand towing caravan and building your layout in that. Getting power to it is not difficult and it's been done before.

Fig. 3/3

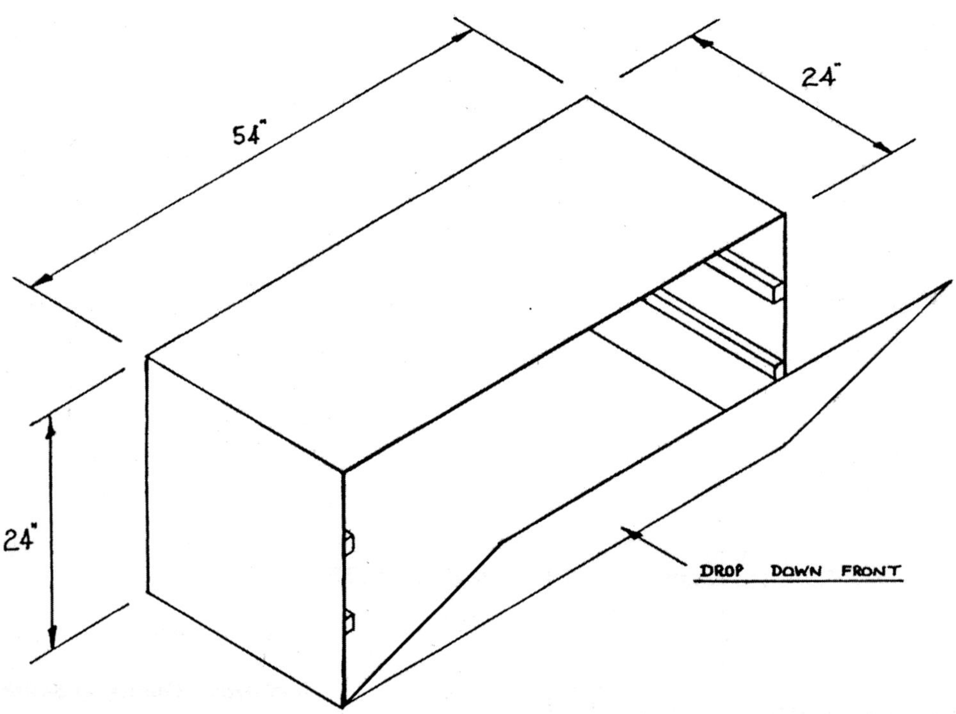

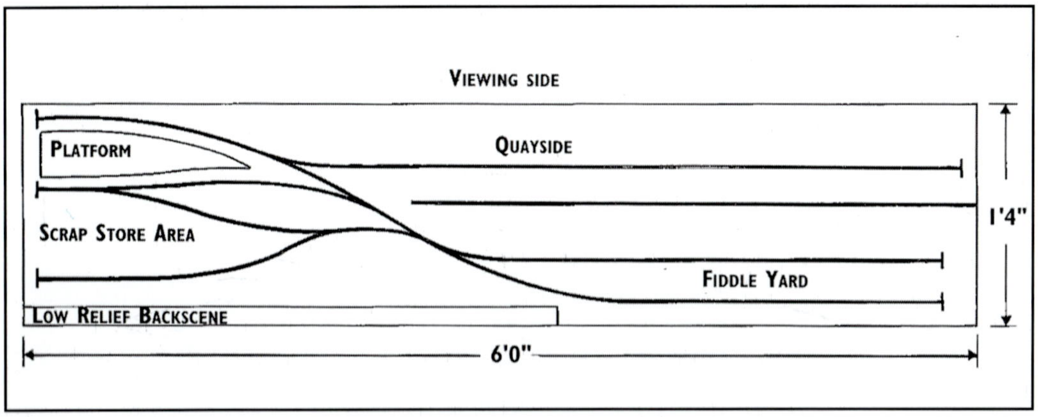

(Left) Fig. 3/4 The track plan of "**The Wharf**". A 7mm N.G. layout which was stored on end in a large built wardrobe together with Justin's 'OO' layout measuring 5ft -6in. x 1ft-4in. called Brookshurts Stabling Point.

Courtesy British Railway Modelling

(Above Left) Photo. 3/3
Two Peco vans stand on The Wharf. The simple Talyllyn Railway card kit used as a backscene shows up well. Note the detail. The man about to climb the ladder, the folded tarpaulins and the people all combine to make a little cameo.

Tony Wright, Courtesy of British Railway Modelling

(Above Right) Photo. 3/4
The Railcar about to leave. The idea of the tarpaulin which covrs up the line motor came from photographs of the I.O.M. Railway. Note the front of the baseboard has been made to look like a wharf side and was made from Faller card "dirted up".

Tony Wright, Courtesy of British Railway Modelling

(Left) Photo. 3/5
"Talyllyn" minus its cab and renumbered No. 8 on this railway stands in the siding behined the scrape dump. The dump shows you should never throw spare bits away.

Tony Wright, Courtesy of British Railway Modelling

CHAPTER FOUR

4. Indoor Locations (In more detail).

Let's start this chapter with stored layouts and possible storage locations. As I said in Chapter 3, I built a layout on an ironing board. The plan is shown in **Figure 4/1 & Photograph 4/1.**This was stored in the garage against the wall. The stock requirements were minimal and the rolling stock was stored in a small box. A small suitcase would do just as well. This layout was quick and easy to set up and dismantle. When shown at exhibitions I and my fellow operator could be in the car and on our way home 10 minutes after the show closed.

I am not the only one to have built a layout on an ironing board. Don Holmes described his N gauge loco depot in the April 1997 issue of BRM **(Figure 4/2 & Photograph 4/2)**. In his article he stated that he had seen a layout called Biginnar Road at the 1994 Shipley MRC's exhibition built by Revd. Dr. Paul Severs to P4 standards. He asked Dr. Severs for permission to use his idea and this was readily given but Dr. Severs pointed out kindly that the real credit for the idea should go to me because I had described my layout in the December 1968 issue of The Model Railway Constructor.

I have also built an 'O' gauge layout on an ironing board! That is slightly misleading in that the ironing board was the support for half of the layout. The other half was bolted to the first part and supported at one end by its own leg **(Figure 4/3 & Photograph 4/3)**. **Photographs 4/4 & 4/5** show two views of Bottrill Street Yard Mk 2, the plan for which is shown on page 27. Ironing boards can be made wider by the use of outriggers which is what I did with this layout **(Figure 4/4 & 4/5)**. The two halves of the layout were stored face to face in the garage. Clearly stored layouts must be properly protected. In my experience, if they are, they can be stored in the garage or the shed without any problem. ***... continued on page 14***

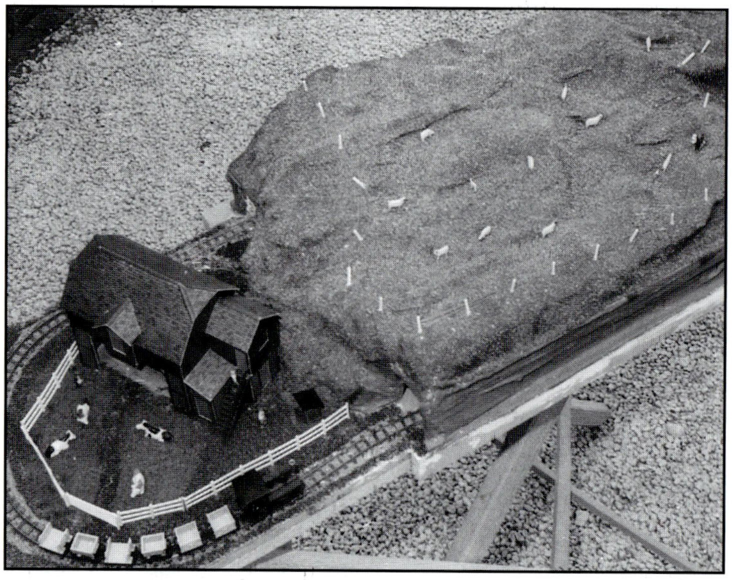

Photo. 4/1. Overall view of my layout on an ironing-board. ***Author***

Fig. 4/1

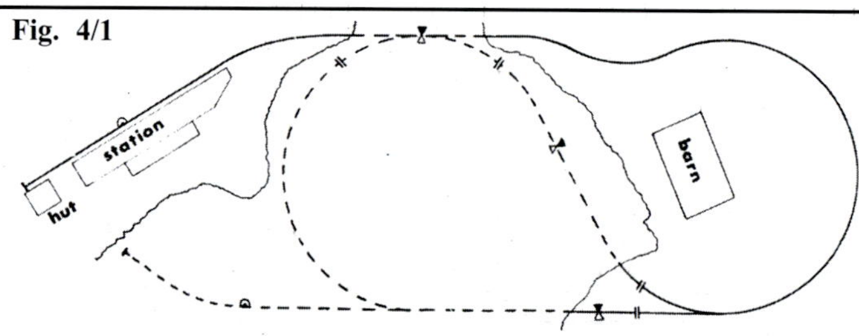

Fig. 4/2 Trainspotters Delight

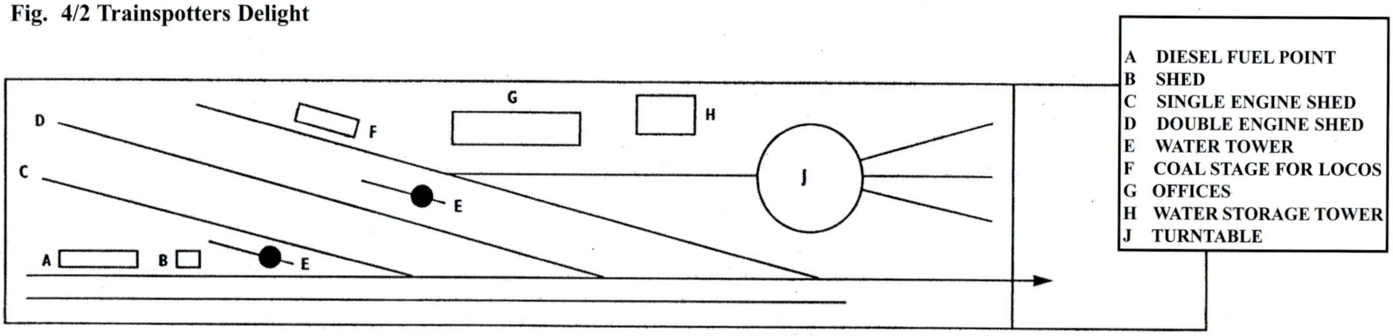

(Left) Photo. 4/2
This view shows the turntable complete with a Southern 'Spam Can' and various other engines on Trainspotters Delight.
Don Holmes/ CourtesyB.R.M.

Fig. 4/3 BOTTRILL STREET YARD

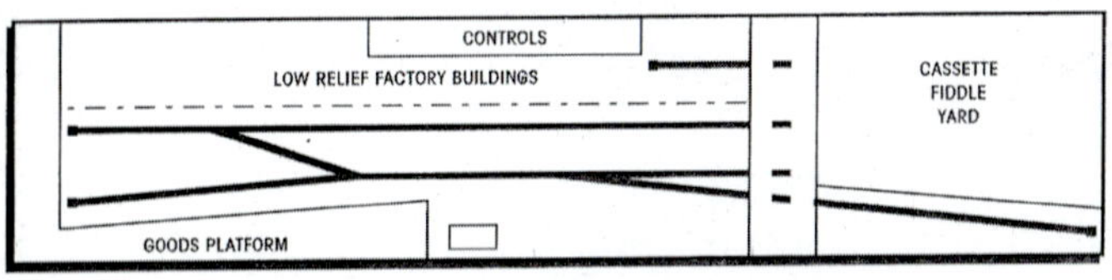

Fig. 4/4 IRONING BOARD WITH OUTRIGGERS

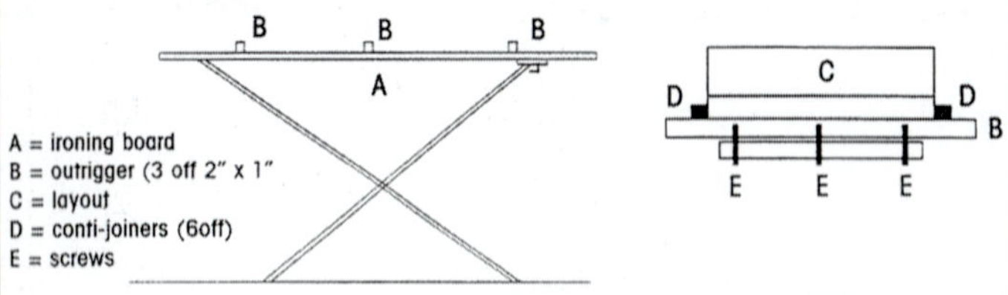

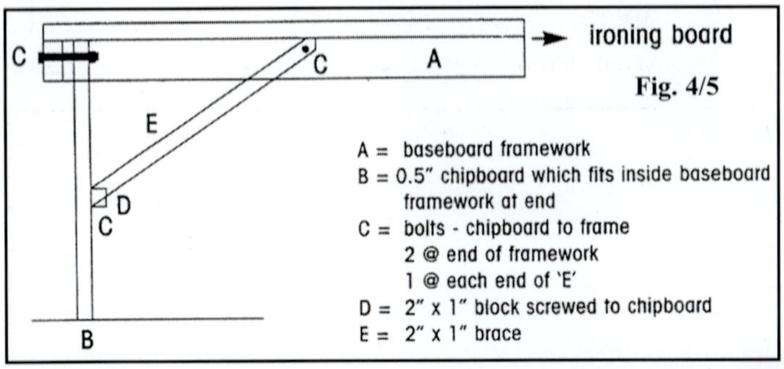

Fig. 4/5

Photo. 4/3
Overall view of Bottrill Street Yard (Mk1) ***Author***

Photo. 4/4
Overall view of Bottrill Street Yard (Mk2). ***Tony Wright***

Photo. 4/5
A further view of Bottrill Street Yard (Mk2) showing the signal box which covers the entrance to the hidden storage sidings.
Tony Wright

continued from page 12 … My current small exhibition layout (The Shed) is stored in the shed (where else!). It has a wooden lid held in position by over-centre catches **(Photograph 4/6)**. I have thought of using the lid as a "roof" supported by 4 pieces of 2" x 1" wood – one at each corner of the layout. The front two supports already have the lighting fascia fixed to them and I envisage black material being fixed with Velcro at the sides of the layout up to the "roof". This would match the material which covers the legs of the layout along the front and round the 2 sides.

The idea of this would be to concentrate the viewer's attention onto the layout making it rather like a theatrical display on a stage.

Some people live in large mobile homes where space is always at an even greater premium that in a small house. About 50 years ago John Charman built an 11ft long 'OO' gauge layout in what was then probably called a "residential caravan". It was split in half and stored face to face in an alcove behind a curtain. It was called Charford and was featured in the Railway Modeller on a number of occasions. If I remember correctly it formed the basis of a larger layout in a shed when John Charman retired from the R.A.F., although on the initial move from the caravan it was simply enlarged by a 2ft 6in. square corner piece. This made a great difference to its appearance and operation. At the same time the fiddleyard was redesigned.

A fellow volunteer on the Talyllyn Railway – Roger Whitehouse – built a layout in a suitcase some years ago called "Michaelstow". Perhaps, more correctly, he built a small portable layout which he transported in a suitcase. The layout (which is of a small Southern Region branchline), its supporting legs, controller, cables etc. have been taken to shows by rail. The stock is carried in a second case. To see Roger erect the layout is extremely interesting and it shows how, with ingenuity, no-one need be without a layout. Roger's measures 9ft x 1ft. Obviously when not in use the suitcase can be stored wherever is convenient at home. **Photographs 4/7 to 4/26** show the layout being assembled at Tywyn M.R.C. 2003 exhibition. (Captions by Roger Whitehouse.)

I have deliberately shown the photographs of Roger assembling "Michaelstow" because I think they are essential to do the layout, and concept behind it, justice.

In Chapter 3 I mentioned 2 books published by the Model Railroader magazine. In the third chapter of the book "Model Rail Roading In Small Spaces" an N gauge layout is described. It is on what is called 'a fold up, roll away base'. It allows you to quickly and easily fold the layout and roll it away to where it is to be stored e.g. in a cupboard or another room.

… continued on page 20

(Right) Photo. 4/6
"The Shed" with its lid on. ***Author***

(Above) Photo. 4/7
A complete 'OO'gauge railway, requiring only a main power supply and (for public exhibition) a table top, preferably 10ft by 1ft 6in. wide.
David Mitchell

(Above Right) Photo. 4/8
Opening the suitcase reveals three baseboards, shortest (back left) an alternative to that on the right where space is limited. Boards are unframed blockboard, for minimum depth. ***David Mitchell***

(Right) Photo. 4/9
Beneath those three are the two principal boards, separated by a loose hardboard sheet from the boxes of tools etc. for basic maintenance, the power unit, and the legs. Domestic use on the floor uses 18in. legs, as seen here. ***David Mitchell***

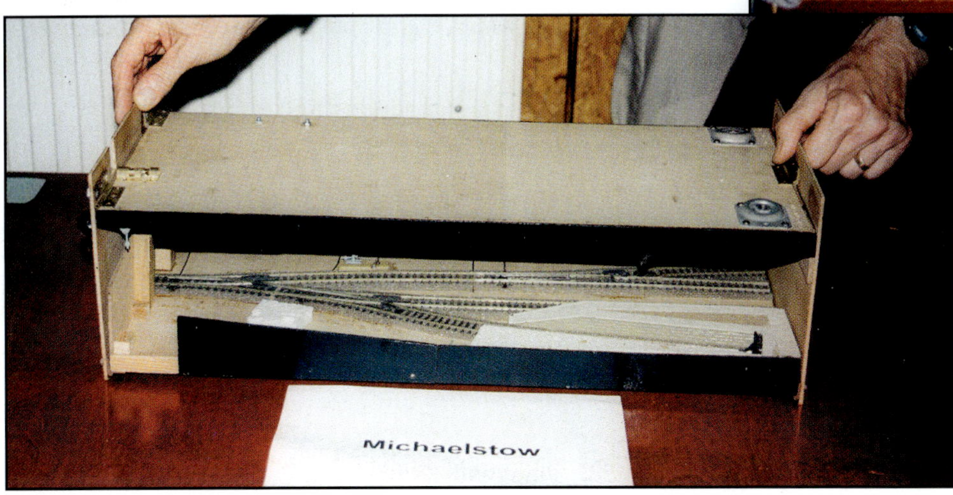

(Left) Photo. 4/10
After removing the tools etc. including shorter legs for exhibitions the main baseboards are lifted out and set down on the table. The screw in leg sockets are seen clearly, as are thin ply spacers which hold the baseboards apart for stowage. ***David Mitchell***

(Left) Photo. 4/11
After screwing in the first set of legs and standing on the table the boards are separated using one leg as a handle. ***David Mitchell***

(Below) Photo. 4/12
The trickiest stage! Holding the handle with one hand, sliding the end spacer sideways will release the upper board allowing it to be pivoted upwards and sideways. ***David Mitchell***

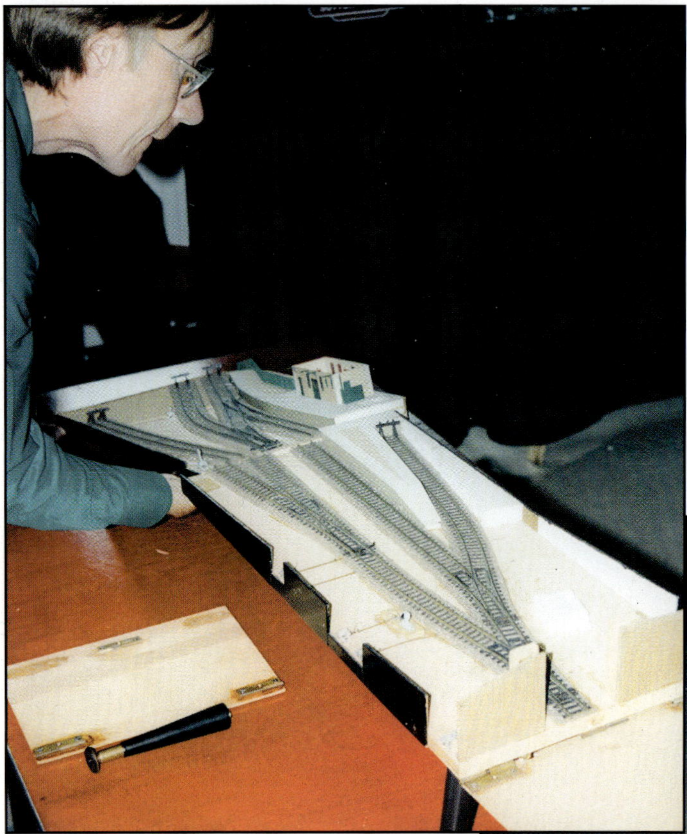

(Above) Photo. 4/13
The second station board, minus handle, can be slid into place by engaging the hinges underneath. ***David Mitchell***

(Right) Photo. 4/14
It is now ready for its legs to be screwed in. ***David Mitchell***

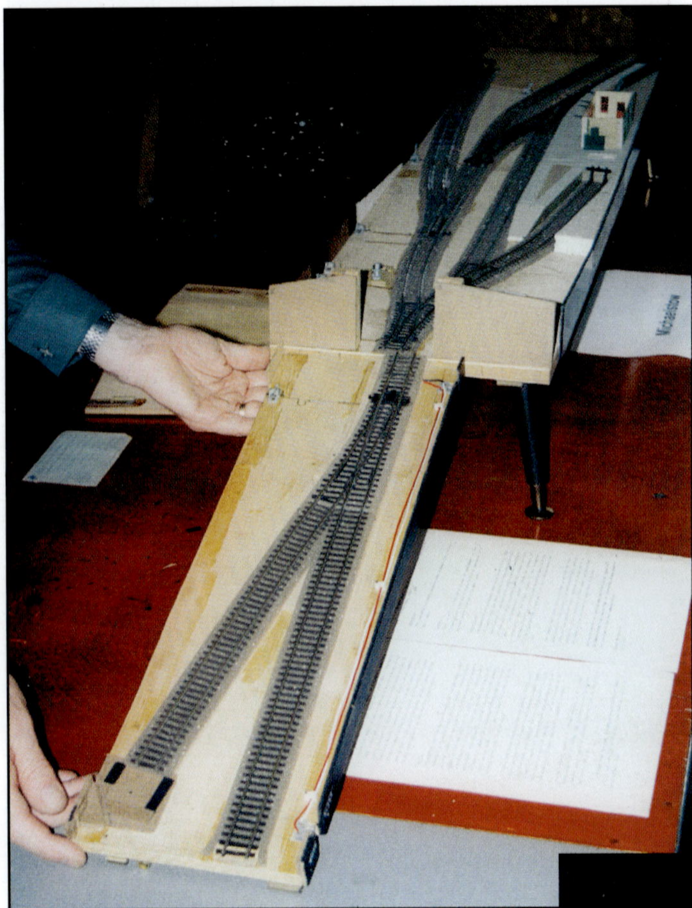

(Above) Photo. 4/15
Sliding the third board into position engaging the hinges underneath. ***David Mitchell***

(Above Right) Photo. 4/16
And finally adding its leg. ***David Mitchell***

(Above) Photo. 4/17
Finally the sector plate and its legs. ***David Mitchell***

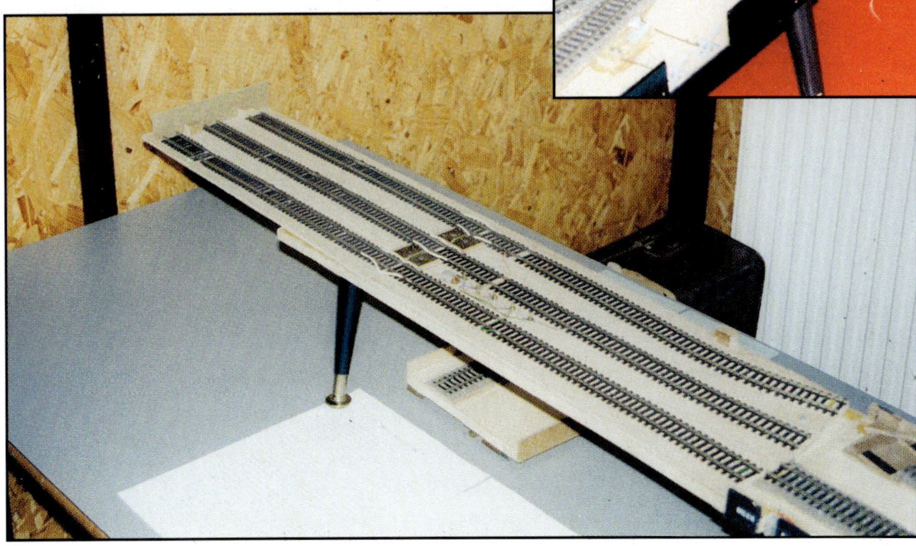

(Left) Photo. 4/18
Leg sockets are fitted to angle the legs for maximum stability. The pivoted sector plate unfolds outwards, giving a layout 9ft long on four 2ft boards. ***David Mitchell***

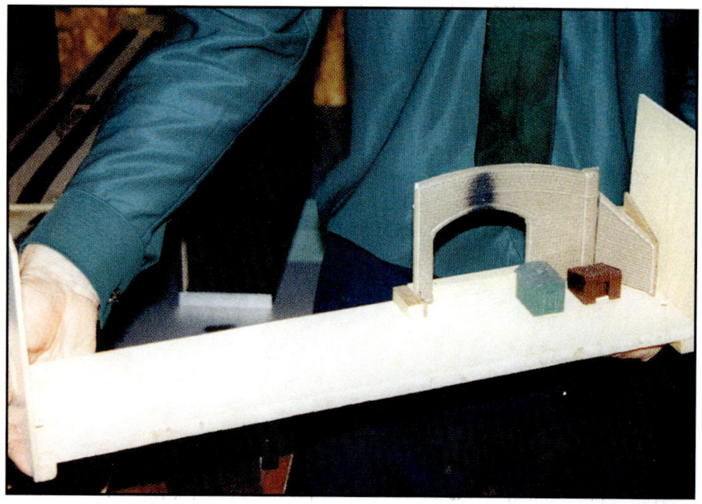

(Left) Photo. 4/19
Electrical connections between boards are made by sliding rail joiners lengthways with long nosed pliers *David Mitchell*

(Above) Photo. 4/20
The scenery park will hold items which have to be removed for storage, including the station lamp hut (which fouls the overbridge in storage) and most of the scenery on the third board. *David Mitchell*

(Below) Photo. 4/21
The bridge screening the 'fiddle yard' and the siding buffer stop are now installed. Eventually the siding will serve an oil depot. *David Mitchell*

(Below Left) Photo. 4/22
The end of the wooden box (held in place by aluminium angles) slides sideways revealing two stock boxes beneath a scenery park.
David Mitchell

(Left) Photo. 4/23
The stock boxes ready for emptying and placing the wagons, coaches etc. onto the sector plate. ***David Mitchell***

(Above) Photo. 4/24
The sector plate with some of the stock. ***David Mitchell***

(Left Middle) Photo. 4/25
The point levers, at the front of the layout, are seen. These must be on the operator's side, so exhibition spectators view from the back and there is no backscene. For the present, extra show boxes have to be used for storing and transporting the incomplete buildings including the station building roof and canopy, signal box and the yet roofless goods shed. ***David Mitchell***

(Below left) Photo. 4/26
Roger sat behind his layout ready for operation! The window ledge made convenient parking for tool boxes. ***David Mitchell***

continued from page 14 ... A similar idea is shown in Chapter 5 of the book where the layout is built on a 'rolling utility cart'. This time the layout does not fold away and is designed to remain horizontal on top of a cupboard on castors. The cupboard would be best utilised for stock and controller/transformer storage as well as, maybe, tools etc.

Turning now to the various rooms in the house. It goes without saying that any layout, which is very visible in the main living area, must be attractive and well presented. When I started railway modelling in the late 1940's, it was commonly thought that you built a layout on a baseboard measuring either 6ft x 4ft or 8ft x 4ft. A layout like that will dominate any room unless the room is very large. Secondly, it is extremely difficult to move; after all it's about the size of a double bed! In any case, by the time you have added buildings and scenery, it will have become extremely heavy, let alone very unwieldy.

If the layout is to be permanently erected and not fill the room, it is much better to build what Cyril Freezer calls a "wall hugging" layout, or to have the layout along one or two walls of the room. Maybe the layout is built on purpose built furniture or on shelving over other items of furniture. Either way, it will be far less intrusive than a 6ft x 4ft baseboard. One well known example of a "wall hugging" layout was built by Dave Howsam and Ron Prattley and described in the Railway Modeller in the 1960's. See **Photograph 4/29**. It was called 'Howley Town'. Two further examples are shown in **Photographs 4/30 & 4/31**. In **Photograph 4/28** you can see Ron Prattley's own layout. The extension at right angles to the bookcase is the lid/top for the layout.

Photograph 4/27 shows Dave Howsam's own layout in the closed position. Surely no-one could object to such a good looking piece of furniture in the lounge!

Another superb example of a layout along one wall of a room ('The Pioneer Valley Railroad') is described in "Small Railroads You Can Build", the Model Railroader book I mentioned earlier. It is in, what our American friends call, the basement. It measures 12ft x 2ft 6ins. and is operated from the front. It rests on purpose built cupboards and shelving and the book has over 20 pages describing the various facets of the layout. A block plan showing the possibilities of a layout along two walls of our lounge is shown in **Figure 4/6**.

(Above) Photo. 4/27 *Ron Prattley*
Dave Howsam's own layout in the closed position.

(Below) Photo. 4/29 *Ron Prattley*
Photographs 4/29, 30, 31, show how effective a layout on purpose built furniture can be. There is plenty of room for storage of magazines, rolling stock or whatever you wish. When the 'lid' is on the layout is fully protected.

(Above) Photo. 4/28 *Ron Prattley*
Ron Prattley's bookcase layout some forty/fifty years ago. The extension to the right is the lid/top for the layout.

(Below) Photo. 4/30 *Ron Prattley*

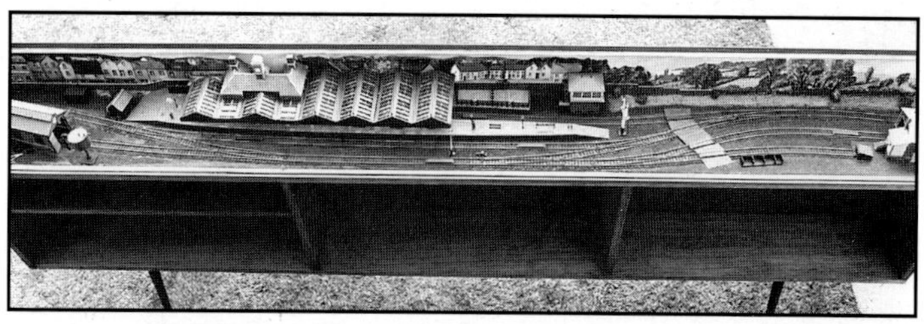

(Below) Photo. 4/31 *Ron Prattley*

Another possibility is the dining room.

When we were first married I built an 'OO'gauge layout in the dining room as mentioned briefly in the previous chapter. It was called 'South Muxton' and was described in the September 1967 issue of the Model Railway Constructor. Looking back now to 1967 I would not be too happy with the standard of the modelling but it was a stage that most of us go through, i.e. flexible track and proprietary stock and scenery. However it met my needs at the time. (**Figure 4/7** and **Photographs 4/32 & 4/33**)

A much better example is 'Gas Street Yard' as built by Mike Hewitt. It was an 'O' gauge layout measuring 10ft x 1ft. (**Photographs 4/34 & 4/35**). The length was dictated by the length of the longest wall in Mike's dining room and the width was kept to 1ft so as to take up as little of the living space as possible. This allowed for the room to be used as a dining room on a daily basis if required. Mike would make the point that it is essential for small layouts to be able to maintain the operator's interest over a long period and I very much agree with him.

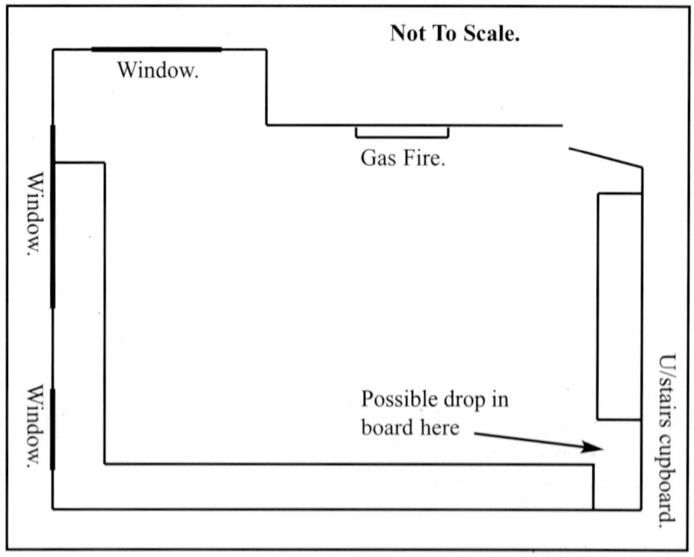

Fig. 4/6 Block plan of our lounge & possibilities for a layout.

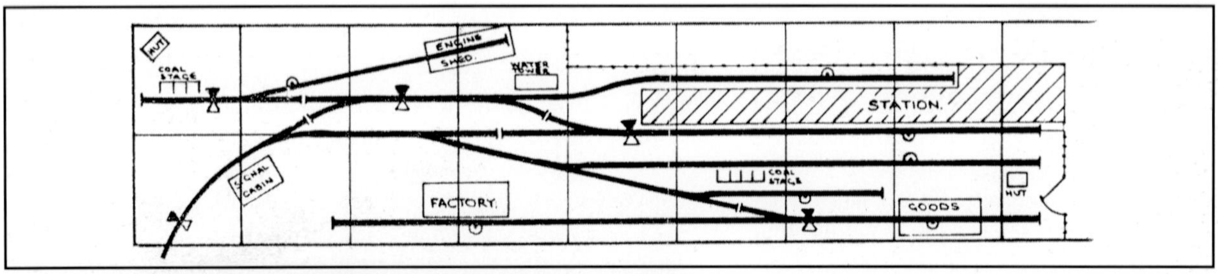

Fig. 4/7
Plan of South Muxton

(Below) Photo. 4/33 A view of South Muxton station signal box, factory and loco shed.
Author

(Above) Photo. 4/32 A view of South Muxton station. All buildings are from proprietary kits. ***Author***

(Below Left & Right) Photo. 4/34 & 4/35 Two view of Gas Street Yard. Particularly note the superb detail. ***Martin Hewitt***

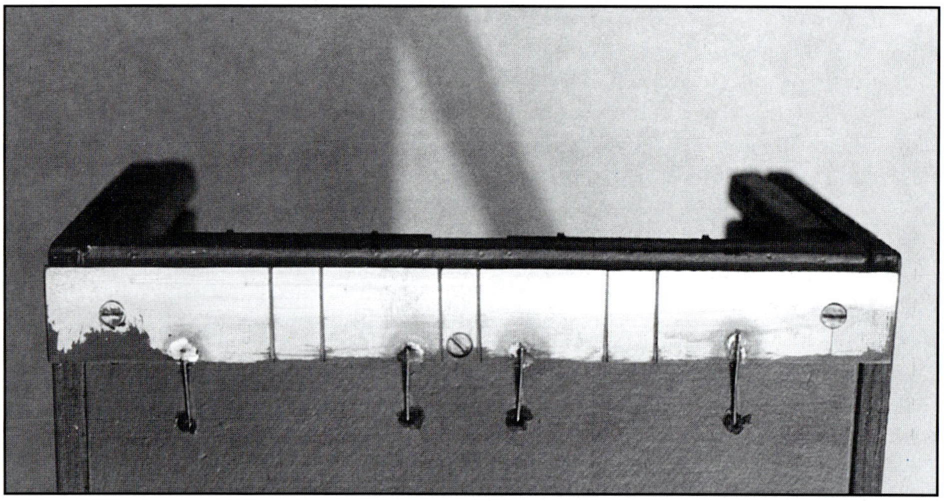

(Left & Above) Photo. 4/36 & 4/37 Two views of ***Mike Hewitt's*** cassettes for Gas Street Yard. ***Martin Hewitt***

Therefore the time spent planning a small layout is time very well spent. I use lining paper which comes in rolls about 22 inches wide and can be bought for between £1 and £1.50 a roll. I then plan the layout full size using points or point templates and a yard or two of flexible track. I also check the operating potential and determine the optimum length of sidings or loops by using the actual rolling stock I intend to use on the layout.

If you have no stock at all you will usually find that retailers are helpful if you tell them why you need the measurements of certain items of rolling stock and locomotives.

Most railway modellers seem to have too much stock for a small layout! However, that's not a problem if you use the cassette type of hidden sidings. There are various ways of making cassettes. They can simply be lengths of MDF with aluminium angle screwed to it. The distance between the horizontal angles is the same gauge of the track. Alternatively you could use Mike Hewitt's design **(Photographs 4/36 & 4/37)**. I use MDF or plywood as a base and lay flexible track and then fit ply or MDF sides. To stop the stock rolling off the ends you can either slide plywood pieces between vertical runners or some people fit wires across. I must admit to preferring something solid for the ends. **(Photograph 4/38)**

However you construct your cassettes, if you have a number of them you can "ring the changes" on the stock that you use.

Layouts such as "Gas Street Yard", which are in a very visible place in the home, should be operated from the front in my opinion. Not only is it easier to work on the layout, it is better to look at from the front.

This raises an issue of whether you intend to exhibit your layout. Do you operate it from the front or the back at shows? Either is possible and there are advocates of both. The choice is yours but, if you intend to operate it from the back, you need to take this into account when planning the control panel. It is quite possible to have the best of both worlds provided you plan for it at the outset i.e. operate from the front at home and the back at exhibitions.

There is a further alternative, depending on the length of the layout, and that is to operate from one end. In that case, I would suggest that a hand-held "walk about" controller is used which simply plugs in on the control panel. This enables the operator to walk in front of the layout for coupling and uncoupling. There are a number of layouts to be seen at exhibitions with the control panel at one end of the layout.

Finally, when considering the dining room, don't forget the possibility of a small layout stored in the room and simply put on the table to operate.

Photo. 4/38 The Author holding one of the cassettes used for Bottrill Street Yard. ***Martin Hewitt***

Some 20 years ago John Allison, a prolific modeller, built such a layout called 'Wolds Wharf'. The baseboards measured 7ft x 1ft 2ins. split 4ft and 3ft. They were joined by permanent hinges raised on blocks and the rear hinge had a small bolt at the back to lock the two boards together when in use. When folded the boards formed a narrow V held together by a strip of hardboard.

Operating a layout like this can set some tricky problems because of the very restricted facilities. You have to know exactly what the capability of the loop and the sidings is. This entails careful measuring of the stock you are going to use. Short wheelbase stock is the order of the day, otherwise buffer locking will result. You need to be a shunting enthusiast to be satisfied with this sort of layout. You could build a layout like this on 'OO' gauge on an ironing board.

A layout like 'Wolds Wharf' could also be run as an American layout or a continental layout with the right stock and maybe different scenic treatment. Equally you could use models of tram locos or Sentinel locos and locate it in a different part of the U.K. The possibilities are many. A description of 'Wolds Wharf' appeared in the August 1982 issue of a magazine called Scale Trains which is no longer published. However, you may find a copy on one of the stands that sells back copies of magazines at exhibitions.

John Allison is a real expert at building portable layouts that store easily and can be readily taken to exhibitions. His 'Oval Ash G Steelworks' was in 'O' gauge and was only 6ft long. **Photograph 4/39** shows the layout folded next to the suitcase with all the stock in it.

John is also a fan of HO gauge as distinct from 'OO' gauge and **Photographs 4/40 & 4/41** show such a layout. The first shows how the layout folds up for storage and transport and the second shows the left hand end of the layout. It is only 4ft 6ins. long. To cap it all, the layout runs on rechargeable batteries! Note the carrying handle on the top of the folded boards and these baseboards are made from balsa wood. You can just see the trade name "Solarbo" on the framework of the small middle board in the photograph.

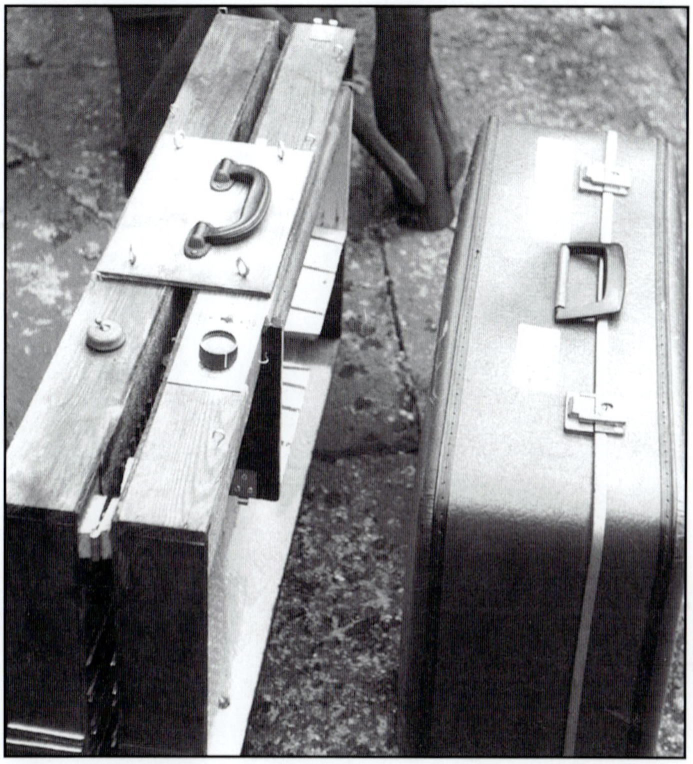

(Above) Photo. 4/39
"Oval Ash G Steelworks" folded next to the suitcase that carries the stock. ***John Allison***

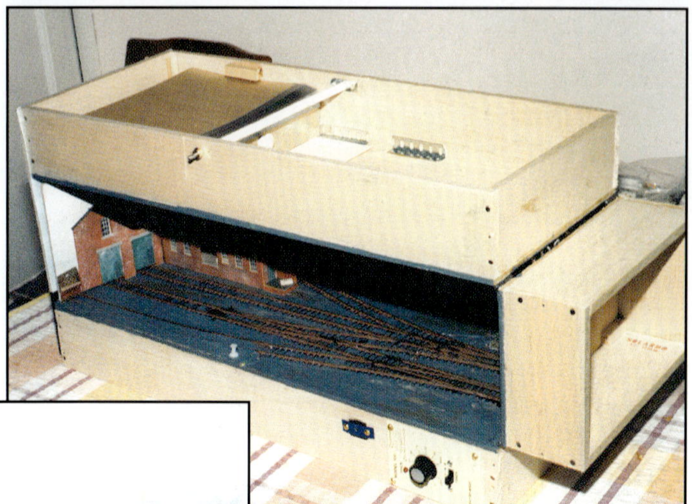

(Above) Photo. 4/40
This view of John Allison's 'HO' gauge layout shows the ingenious way John's design of the boards protects the layout when being transported to exhibitions and being stored. ***John Allison***

(Left) Photo. 4/41
This shows how you can give the impression of tracks going into a low relief building. ***John Allison***

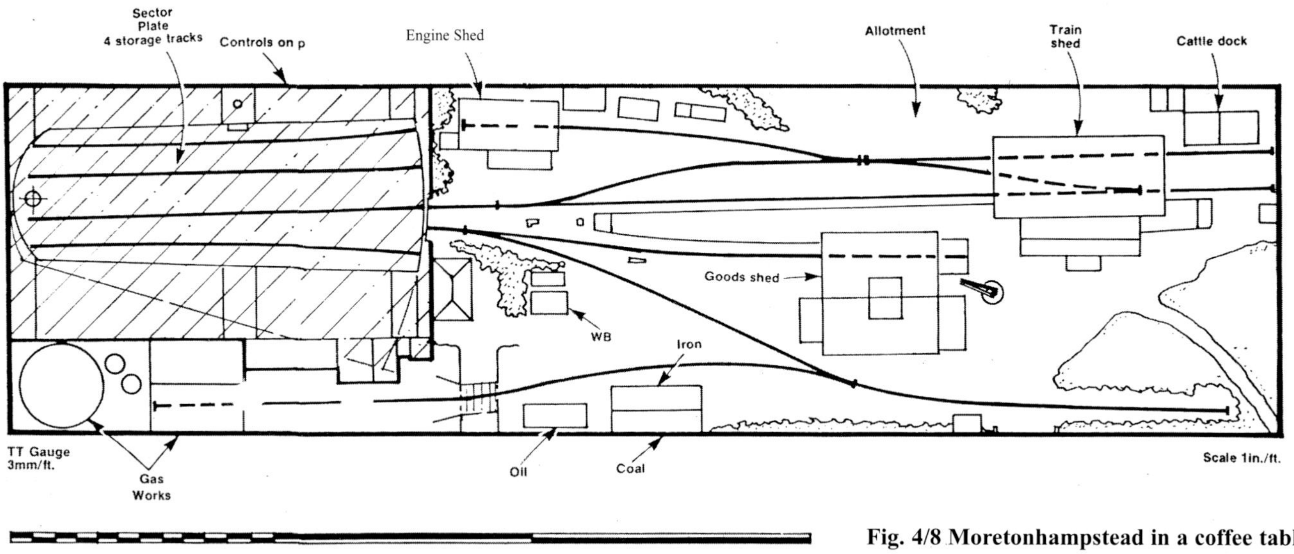

Fig. 4/8 Moretonhampstead in a coffee table.

We come now to the lounge.

In many houses there is one large room described as a lounge/diner. Perhaps the simplest way of having a layout in the lounge is to make a feature of it but keep it portable.

A number of people have built layouts in coffee tables. Some years ago a modeller called Ray Arnold built one in 3mm scale called Moretonhampstead **(Figure 4/8)**. Such layouts can be supported at "coffee table level" by commercially available tapered table legs which screw into "screw plates" on the underside like those used by Roger Whitehouse on "Michaelstow"

Our American friends also build layouts to be seen permanently in the lounge. Two very good examples are shown in the book I previously mentioned "Model Railroading in Small Spaces".

One example is an N gauge layout built in what is called an "end table". It can be used as a table and the layout is under an acrylic plastic top which can be lifted off for access. There is a drawer under the table which can be used for the controls and/or stock storage.

The other example is another N gauge layout built on a "bar top". It measures 5ft x 2ft and stands on two barrels! There is a glass lid which lifts up for access to the layout. The lid is built into a wooden frame which is hinged along the back of the layout box. The layout and the bar are used regularly!

If anyone is considering building a layout in a small space, I would really recommend this book. It is full of very good ideas which have actually been put into practice. The book is published by Kalmbach Books and is available in the U.K. I obtained my copy from "Totally Trains" at an exhibition a few years ago and it cost £12 then. It was first published in 1998 and the author is Mat Chibbaro. The price in dollars was $16.95.

Some 20 or 30 years ago it was the 'in thing' to build layouts in what were described as 'odd places'. One such place was in the case of an old T.V. set. At that time the screens were not as large as you can get today so perhaps the time is right to have another look.

For obvious reasons of space, such layouts are usually built in N gauge or '009' but the advent of Z gauge some years ago adds another possibility. It has to be said that such layouts are "gimmicky" but, when they were in fashion they were "crowd pullers" at exhibitions simply because they were gimmicky and different. So, if you are really stuck for space why not resurrect the idea.

The late Martin Brent, whose untimely death from leukaemia deprived the model railway world of a superb modeller, described such a layout in the June 1981 issue of Model Railways (no longer published). The diagram in **Figure 4/9** shows the basic idea.

In the last chapter I mentioned a layout built by Gordon Gravett. The idea of the layout (described in the January 1993 Railway Modeller) was to make a diorama which was small enough to be stored and displayed, and may be operated, on a shelf. Such is the standard of Gordon's modelling that this layout could grace any lounge. When operating he adds a fiddle yard at each end (see plan in **Figure 4/10**). These could easily be stored with the stock in a cupboard or chest of drawers. In fact the layout itself could stand on the chest of drawers. The size of the diorama itself is 4ft x 1ft 6ins. and it is made of 4mm plywood braced with 12mm x 12mm stripwood where necessary. The front is protected by Perspex sliding "doors" and integral lighting is built in. The inside of the diorama is made mainly of 4mm x 6mm plywood. The Perspex sliding doors run in plastic tracks fitted top and bottom at the front of the cabinet.

You used to be able to see this layout "in the flesh" in the Narrow Gauge Railway Museum of the Talyllyn Railway in Tywyn as Gordon kindly donated it to the Talyllyn Railway when he stopped exhibiting it. I say "used to" because the Museum is temporarily closed whilst Wharf Station is re-developed.

Whilst talking about the Talyllyn Railway, the Railway Modeller of February 1980 describes another layout built in a similar display case to Gordon Gravett's. This time it is 009 and features the Dolgoch Falls viaduct and station. Behind the back scene there is a loop and therefore a limited "round and round" service can be run. The fascia board incorporates a strip light. It is suggested in the article that, if the display case is really attractive and a really good model is mounted inside, any opposition to "living room modelling" may well disappear.

Fig. 4/9

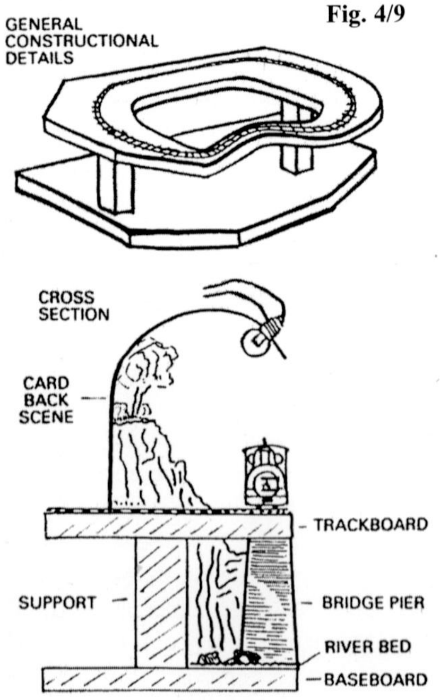

Fig. 4/10

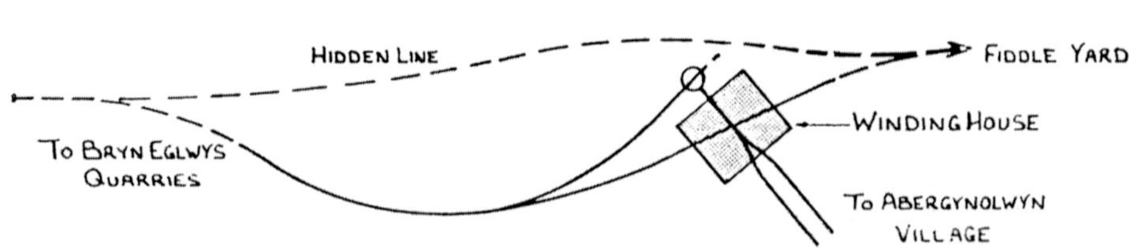

Fig. 4/11 ARCADIA

Martin Brent once built an exhibition layout called "Winchelsea Road" which was described in Model Railways, July 1980. It was 20ft long and was erected in the lounge on occasions. The only problem was that the lounge was only 19ft 10ins. long! The solution to this was to open the window. I suggest that this is most definitely NOT the way to harmony in the household..

Martin therefore dismantled the layout, and eventually bought a garden shed. However, in the meantime, he built a small layout which, in his case, fitted in an alcove in one of the bedrooms. Many houses have alcoves in the lounge, (or dining room), often either side of the fireplace.

Martin's standard of modelling was superb and one of his layouts would have been an asset to any living area.

This layout was called "Arcadia" and was 7ft 5ins. long x 1ft 6ins. wide. It was built on 2 baseboards, each 44fi ins. long; the scenic section was 5ft 2ins. long. A plan is shown in **Figure 4/11.**

The layout was described in the March 1981 edition of Model Railways. Of necessity, the operation on such a layout will consist mainly of shunting but this often creates a great deal of operational interest. Layouts like "Arcadia" are also capable of being extended if additional space presents itself or for exhibition purposes. The latter was true in Martin's case and "Arcadia" formed the basis of "Rye Harbour" which was described in Model Railways over a period from November 1982 to April 1983 and appeared at many exhibitions.

Iain Rice's book 'An Approach to Model Railway Layout Design', subtitled 'Finescale in Small Spaces' (published by Wild Swan Publications Ltd) gives a number of very good ideas for use in the living area of a home. It is a very worthwhile investment if you want to build a small layout. One in particular would be very suitable for an alcove. It is called "Bankfoot" and measures 7ft 6ins. x 1ft 9ins. and is operated from the front. The inspiration for the layout is the Clee Hill Granite Tramway which was a 4fi mile mineral line operated by the LNWR not far from Ludlow.

Also described in that book is a layout for what Iain calls a "deep corridor". He envisages the supporting structure being on wheels so that it can be moved out from the walls for decorating or access for cleaning. He suggests the use of drapes below the layout to form the "stage" to minimise the impact on the room in which it is located. Iain has written another book entitled 'Designs for Urban Layouts' (published by Atlantic Publishers) and this too is well worth purchasing.

One very well known layout called "Inglenook Sidings" was called that because it occupied the space beside the chimney-breast, 'the inglenook'. It was built by Alan Wright of the Manchester MRS and was first exhibited at the Manchester Exhibition in 1979 and attracted a great deal of interest. Since then it has been copied many times in various scales, small and large. It is in 'OO' scale and measures 4ft x 1ft and is totally self-contained apart from the 6ft wander lead for the controller. This layout is a "shunter's paradise" using a small loco, eight wagons and a brake van. It represents a small English country goods yard.

Alan uses what he calls the "Tiddly wink computer". Eight tiddly winks each have the name of a wagon marked on them. You then shake them up and lay 5 down in random order. You then shunt the train to make it up into that order. Not as simple as it sounds as the brake van cannot be moved off the main line.

A new version of Inglenook sidings was described in the Railway Modeller of December 1992 as it was shown at the Manchester Exhibition at that time.

Finally, when thinking about living space, don't forget that, if the 'household authorities' agree, you can put a layout on top of existing furniture. I know of at least 2 layouts that live on top of the piano.

Let's move on now to the possibility of having a layout in a bedroom.

The first question to ask is "can it go in a spare bedroom?" Obviously if you have a spare bedroom which can be given over entirely to model railways that is utopia! Some many people might say it is not utopia because the spare bedroom is nearly always the smallest one and therefore unlikely to be large enough.

However, you may well have to accept that and design a layout to suit. Personally, I would much rather have a small room which is entirely given over to model railways than a larger one which has to have the usual bedroom furniture in it too.

Let's deal with 'utopia' first and with those examples of my own.

One undoubted advantage of being a vicar is that you live in a 4 bedroomed house. One further advantage in my case was that our 2 sons had grown up, and not all the bedrooms were required, even though they each had a bedroom which was "theirs". Therefore I was able to have the smallest bedroom for model railway use. By this I mean that the room accommodated not just a layout but my work bench and storage as well. The only work that was done elsewhere was "dirty work" such as sawing wood, painting, etc. which was done either in the garage or outside on a fine day.

At Wyken the room was set out as shown in **Figure 4/12.**

The layout I built there was "Wyken Yard" which was in 'O' gauge and on three boards, each 38ins. x 22ins. (See **Figure 4/13** and **Photographs 4/42 & 4/43**) Whilst at Wyken I also built my Small M.P.D. (BRM October 1995) which measured all of 45ins. x 22ins. in 'O' gauge! This could be erected in the layout room when required on its own purpose built support frame which was made for me by a parishioner to his own design. The plan of the Small M.P.D. is shown on the control panel **Photograph 4/47** and the support frame is shown in **Photograph 4/48**. The traverser is in the area on the panel marked 'Motor' and 'Track' The switch put power to the traverser motor or the tracks.

When I moved to Nuneaton of course the spare bedroom was a different size! However, I was still able to accommodate layouts, workbench and storage. **(Figure 4/15)** I was able to double the size of the Small M.P.D. which I really enjoyed operating (See **Photographs 4/45 & 4/46**) and built another layout called "Bottrill Street Yard" (BRM June 1998) and the room was not overcrowded. **Photograph 4/44** shows my elder son, Justin, operating Bottrill Steet Yard at the N.E.C. On retirement, the luxury of a 4 bedroomed vicarage had to go and we now live in a 3 bedroomed cottage about 100 yards from Pendre Works on the Talyllyn Railway where Celia and I are volunteers, as is our elder son.

Now our needs are different. Justin comes to stay when working on the T.R. and friends and relatives come for holidays. Nevertheless I am still able to use the spare bedroom as the place to locate my latest layout, workbench and computer work station.

A plan of that bedroom is shown on Page 7 and the plan of my latest layout is shown in **Figure 4/14**. The optional scenic boards are stored in the shed in the garden in a wooden box specially made to suit. This also

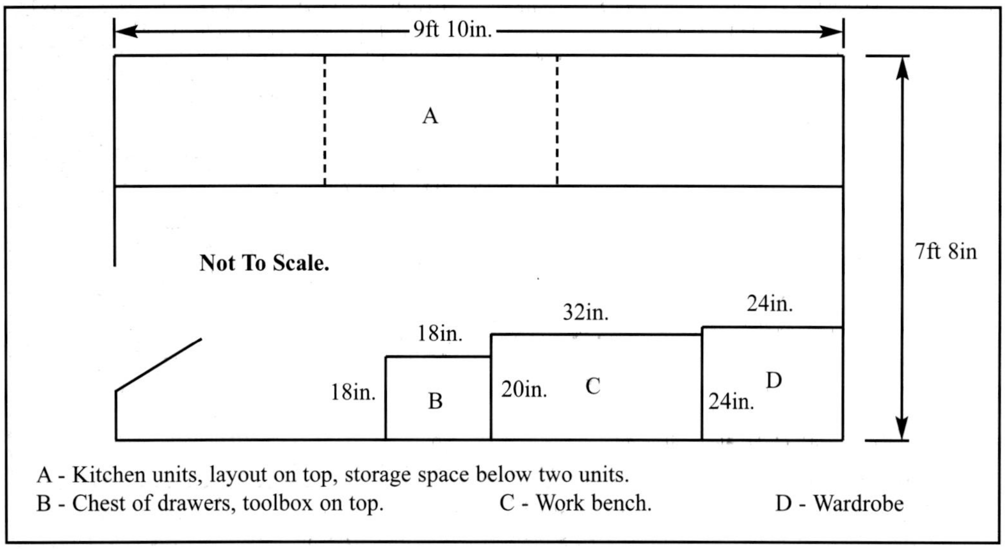

(Left) Fig. 4/12
Layout of my modelling room at Wyken.

(Below) Fig. 4/13
Plan of Wyken Yard layout.

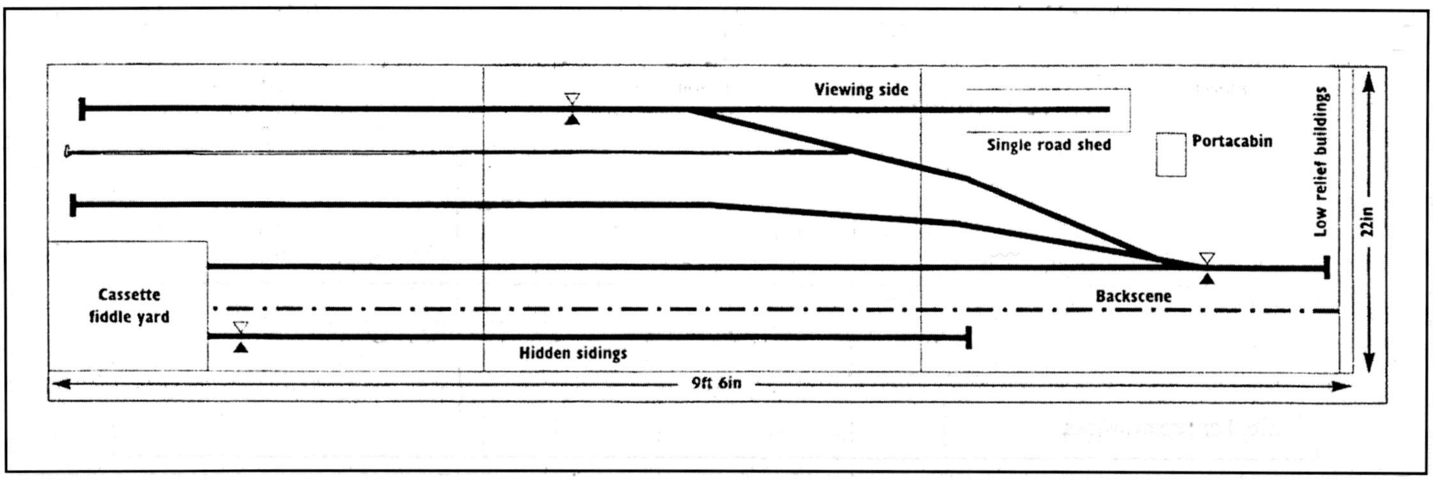

Photo. 4/42 A view of the Morris Minor specialist with the preserved Coventry bus on my Wyken Yard layout. ***Martin Hewitt***

Photo. 4/43 A more general view across 'Wyken Yard'. ***Martin Hewitt***

enables those boards to be easily transported to shows. To avoid any confusion I should explain that this new layout is called "Bottrill Street Yard Mk. 2" simply because I had a professionally made sign for the original layout of the same name and it was too good to discard! **Photographs 4/4 & 4/5** can be seen on pages 13 & 14.

The main bedroom in our cottage now has a built in wardrobe but, prior to that, there was an alcove which would have been very suitable for a layout. (See **Figure 4/16** for a block-plan of the room.)

Likewise there would be room for a layout in the second bedroom which could be either self-contained or have a fiddle yard which would be stored underneath the layout and brought out for use. Depending on the design for the layout, that extra board could go out onto the landing. (See **Figure 4/17** for possible block-plans.)

It may be that you want to use the bedroom to store and operate the layout but not leave it permanently erected. This has been done using a wardrobe. In Model Railways December 1980 G.A.Evens put forward an ingenious idea for building a layout stored in a wardrobe approximately 6ft high by 2ft wide by 13 ins. deep. The diagram which appeared in the magazine is shown in **Figure 4/18**. I could not ask Mr. Evens' permission to reproduce this because the magazine ceased to be produced a good number of years ago. However, if he reads this, "Thank you very much Mr. Evens".

Photo. 4/44 Justin operating 'Bottrill Street Yard' (MK1) at the National Model Railway Exhibition at the NEC. ***Author***

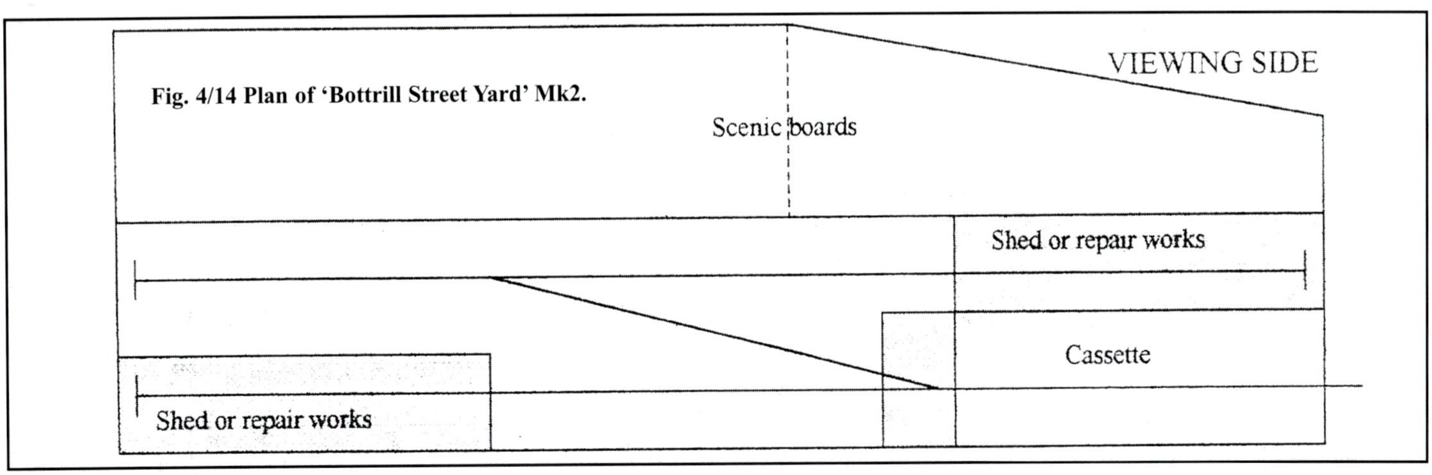

Fig. 4/14 Plan of 'Bottrill Street Yard' Mk2.

Photo. 4/45 The extension to the small MPD. ***Author***

Photo. 4/46 The small MPD as originally built with part of the extension. ***Author***

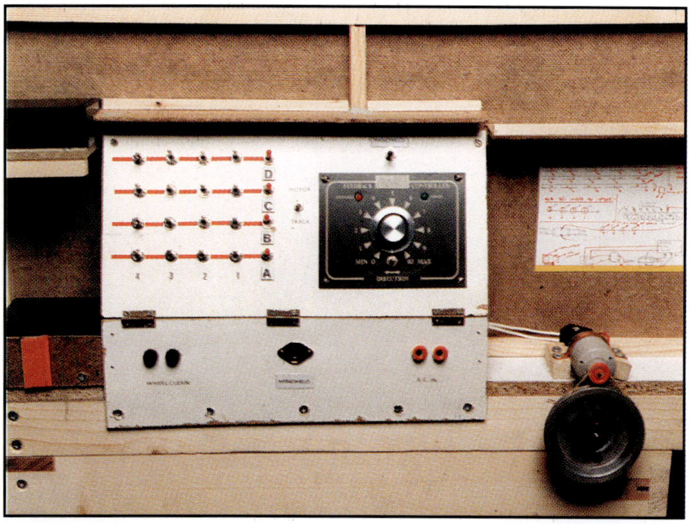

Photo. 4/47 A view of the 'Small MPD' control panel which shows the simple track plan. The motor and pulley drive the threaded rod for the traverser, connected by an elastic band. ***Martin Hewitt***

Photo. 4/48 A view of my layout support frame. It was originally designed and built by a parishioner for 'Small MPD' but is used on other layouts. ***Martin Hewitt***

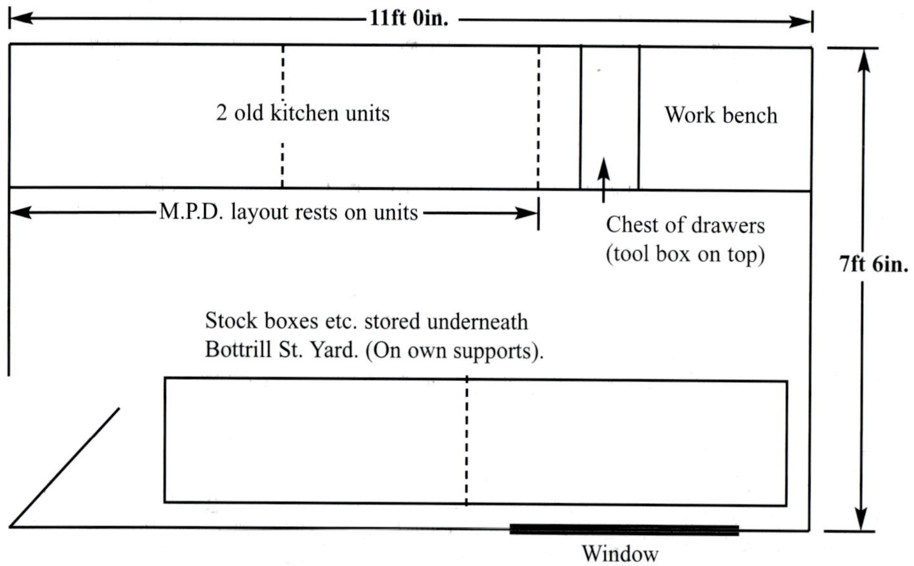

Fig. 4/15 Plan of Modelling Room in Nuneaton.

Note: Layouts can be moved around e.g. Bottrill St. Yd. can be crated face to face and stored in garage. MPD can be moved to where Bottrill St.Yd. is and space is created to work on a new layout. Alternatively, MPD could stay put and a new layout be worked on where Bottrill Street Yard is.

Not To Scale.

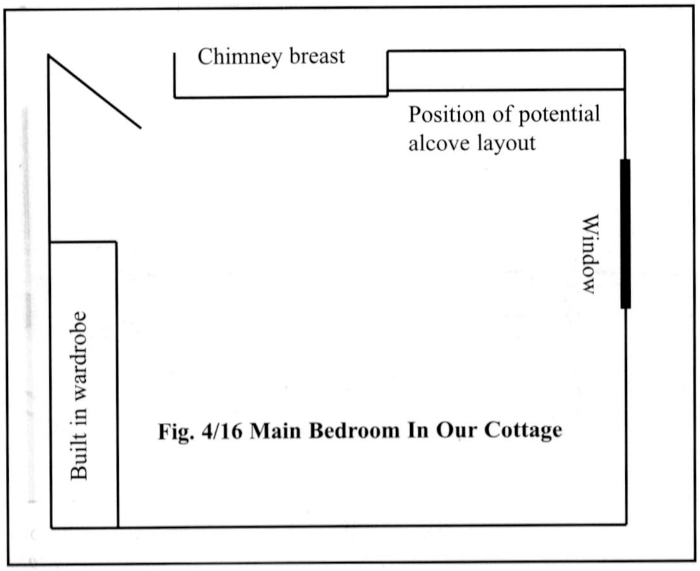

Fig. 4/16 Main Bedroom In Our Cottage

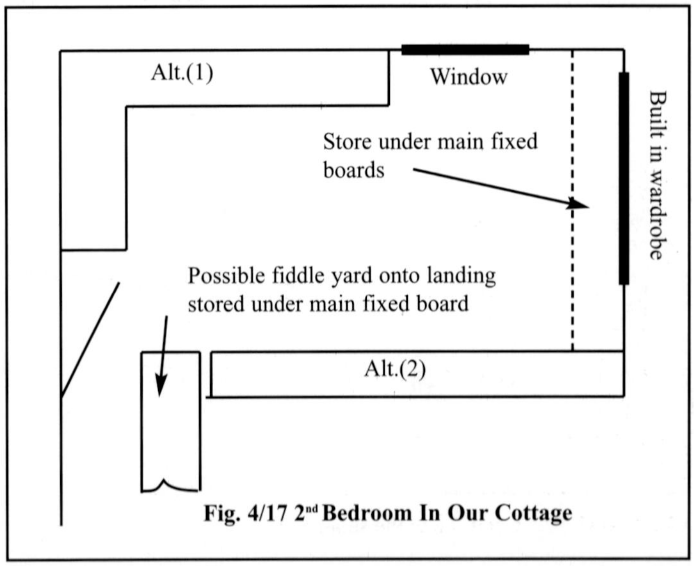

Fig. 4/17 2nd Bedroom In Our Cottage

Fig. 4/18

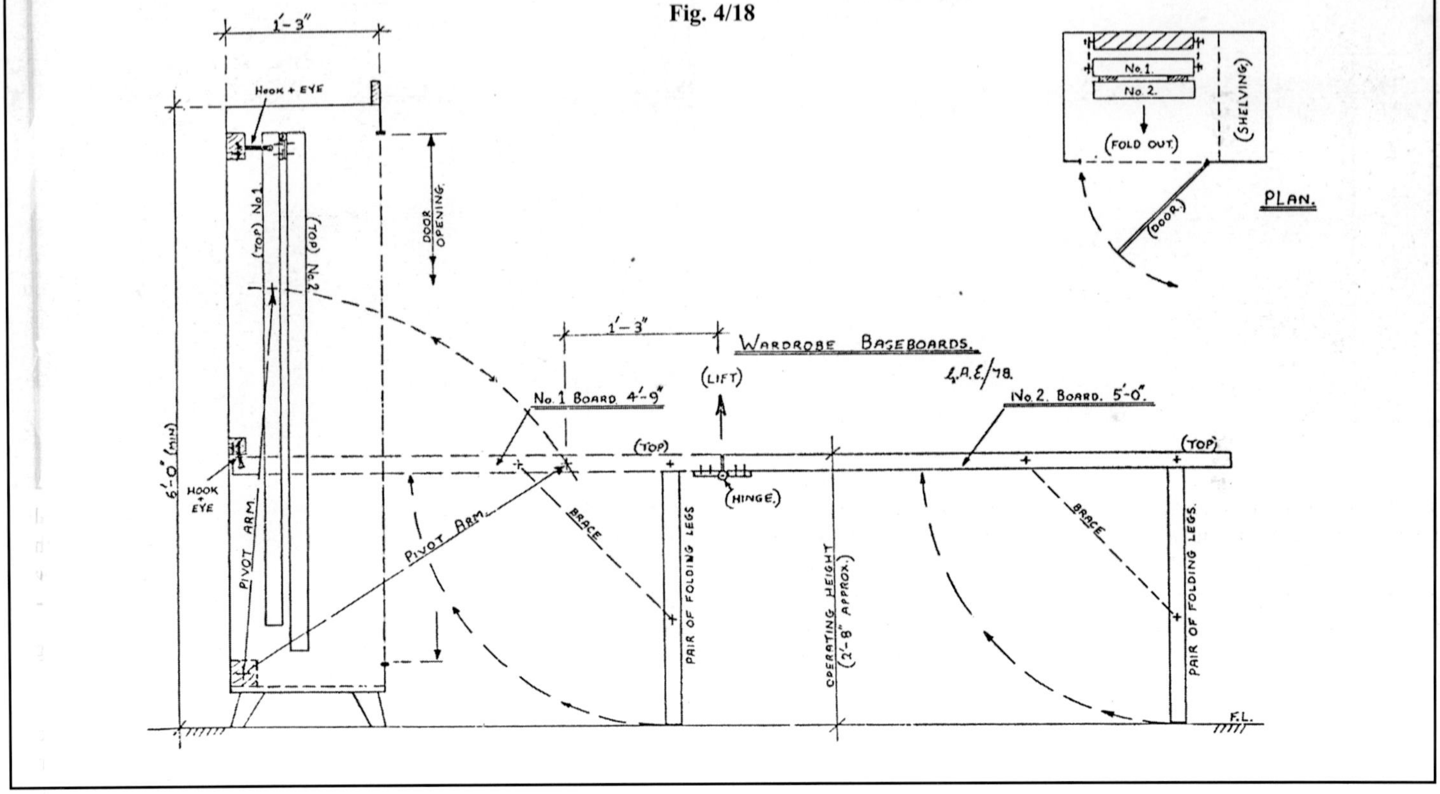

If you have room in the bedroom for a permanent layout such as a "wall hugger" all well and good. Malcolm Crawley describes a layout which would fit along two walls of a bedroom in BRM 1998. It is called "Thompsons End" and is built on bookcases, the tops of which measure 7ft x 15ins. and it is L shaped. The plan is shown in **Figure 4/19** and **Photographs 4/49 & 4/50.** It could easily be shortened by making the station a terminus. Such a layout can be built with exhibitions in mind and made easily transportable. If I was building it, I would design an exhibition carrying frame from the outset but that's probably because I design all my layouts with exhibitions in mind.

Another layout that could be accommodated in a bedroom is "Calder Hebble Yard" an 'O' gauge layout built by Ralph Nuttall, measuring 8ft x 18ins. I say that because, in his article in BRM March 2002, he says the baseboards he used were those previously used by his son for his 'OO' gauge layout in his bedroom. The plan is shown in **Figure 4/20** and, having seen it at one of the Festivals of British Railway Modelling at Doncaster, I would be very happy to have it in one of my bedrooms if it could be fitted in. (See **Photograph 4/51**)

Subsequently Ralph sold this layout to Geoff Silcock who rebuilt it and it can now be seen at shows as "Canal Sidings and Albert Street E.20". It now measures 9ft 6ins. x 1ft 9ins. The plan is shown in **Figure 4/21 & Photograph 4/52.**

Fig. 4/19 Plan of Thompson's End *Courtesy B.R.M.*

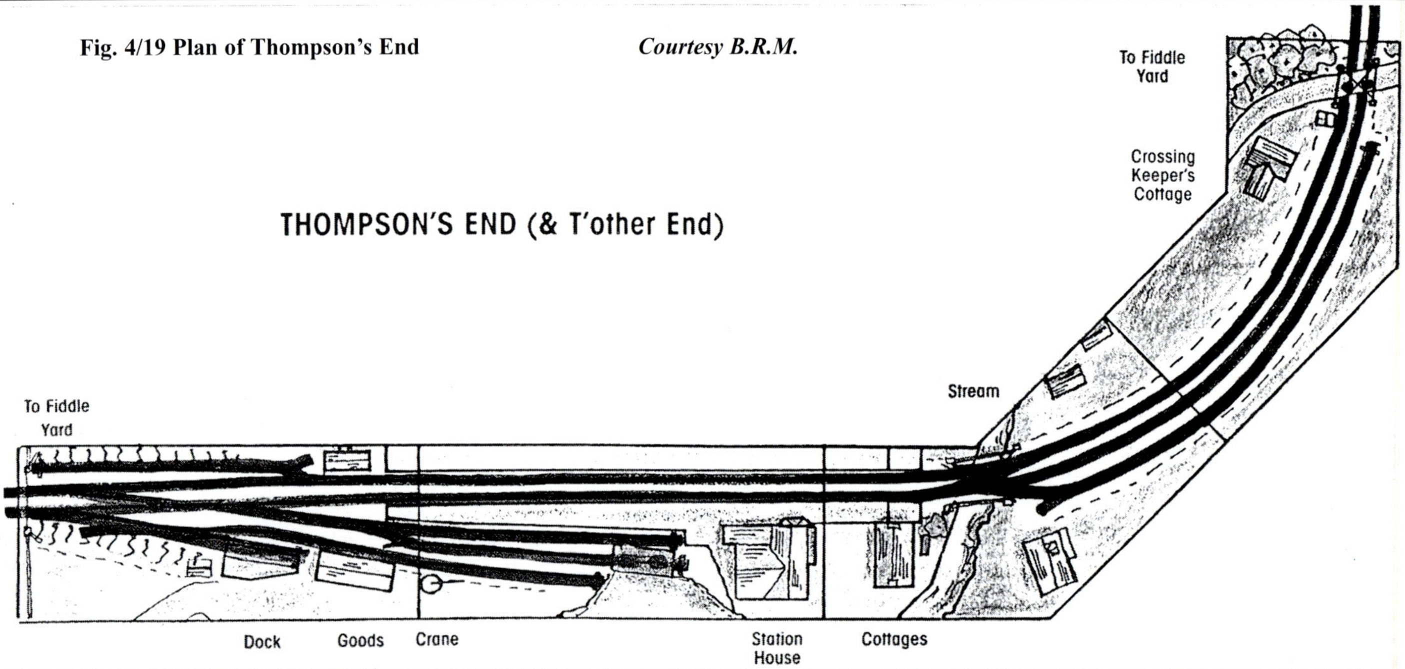

Photo. 4/49 *Tony Wright Courtesy B.R.M.*
A breakdown train is seen passing the relief road on the left of the running tracks.

Photo. 4/50 *Tony Wright Courtesy B.R.M.*
A view looking towards the over bridge which disguises the entrance to the fiddle yard.

Fig. 4/20 Plan of Calder Hebble Yard. *Courtesy B.R.M.*

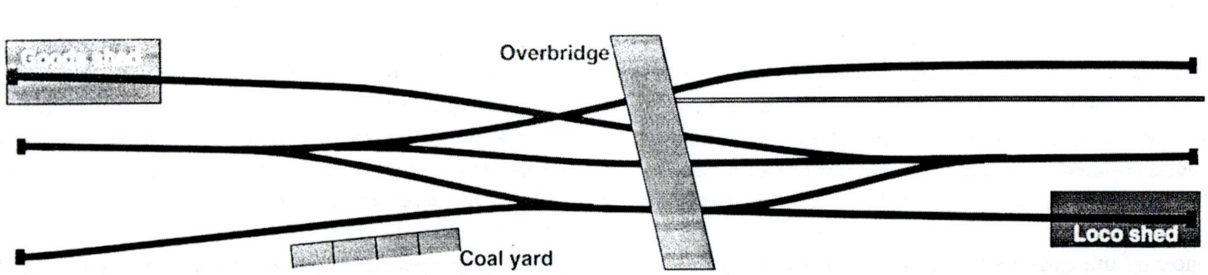

Photo. 4/51 ***Ralph Nuttall/Courtesy B.R.M.***
A view of Calder Hebble Yard as it was originally with the engine emerging from the hidden storage sidings.

Photo. 4/52 ***Geoff Silcock/ Courtesy B.R.M.***
In this view the layout is now under the guise of Canal Sidings & Albert Street E.20.

Fig. 4/21 Plan of Canal Sidings & Albert Street E.20 ***Courtesy B.R.M.***

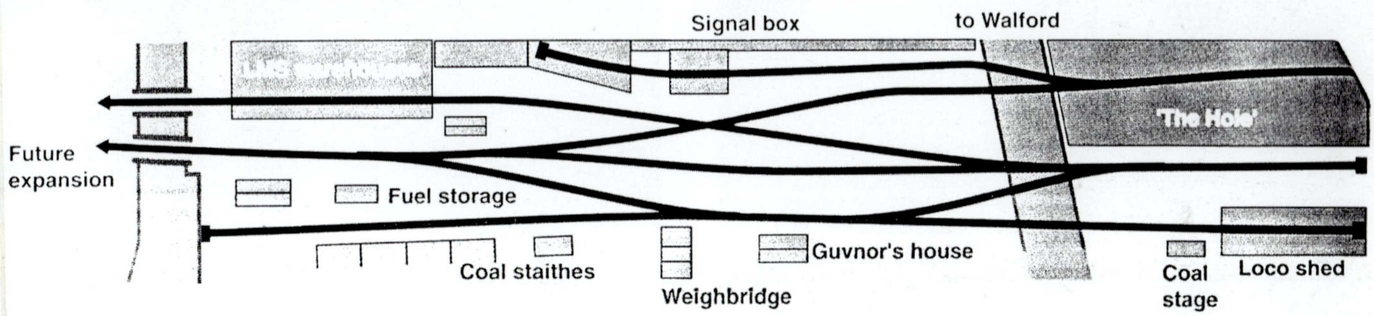

Let's now consider the loft. Utopia as far as roof space is concerned is a house built at the beginning of the 20th Century with rooms in a hip-shaped roof. One of my colleagues had a vicarage with two good sized rooms in the attic with dormer windows and either of those rooms would have been superb for a railway room, not least because you have no access problems, noise problems or height problems because they were designed as rooms at the outset. If you can find such a house, you are lucky indeed!

The essential decision that has to be made if using the loft is whether the layout is to be permanent or taken to exhibitions. This affects the size of access hatch. Personally, I would always make the access as large as possible even if the layout was to be permanent because this makes it easier to get raw materials for baseboards into the loft. Secondly, as I said earlier in this book, you never know if you are going to have to move house. Therefore, it is extremely advisable to build a permanent loft layout in sections anyway to cover that eventuality.

I know of one case here in Tywyn where the layout was NOT built so that it could be easily moved. As a result, the house has changed hands twice now and the layout in the loft has had to be a selling feature. However, that has been advantageous to the Tywyn M.R.C. as the last owner and the present one are members!

Of course, if you have a suitable loft, you can locate your workbench there too and store rolling stock, unmade kits and tools.

One well known 7mm layout which is regularly exhibited is Nigel Bowyer's "Napier Street". (See **Figure 4/22** and **Photographs 4/53 & 4/54)** This is kept in the loft.

The final possibility within the house itself is a basement which is not all that common in this country compared with, say, the U.S.A. and, in my experience, the access stairs are usually narrow and steeper than ordinary ones.

Nevertheless, if there is any tendencey to dampness and it can be overcome, the advantages are obvious and our American friends, whose homes are not designed in the same way as those in the U.K., often make use of the basement for their layout. I have visited the U.S.A. once and was fortunate to see such a layout. Examples can regularly be seen in the Model Railroader magazine which is readily obtainable in this country.

Fig. 4/22 Plan of Napier Street *Courtesy B.R.M.*

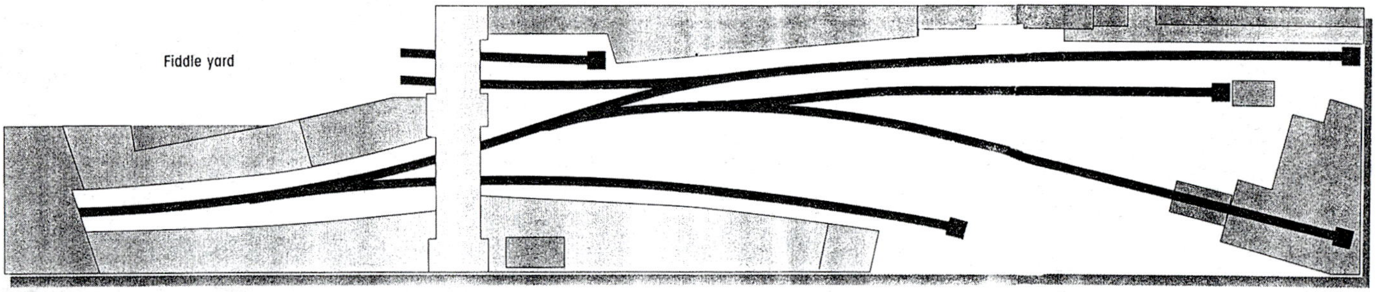

Photo. 4/53 *Tony Wright Courtesy B.R.M*
A view taken from underneath the over bridge which spans the entry from the fiddle yard and the shunting neck. Note the attention to detail like the point lever and roding and the state of the track with it's vegetation growing at the sides.

Photo. 4/54 *Tony Wright Courtesy B.R.M.*
Again note the detail. The shunters pole on the front steps of the Class 08 as well as the lamp on the running plate. Also at the side can be seen the coal pile together with it's weighing scales and sacks.

5. Outdoor Locations.

The most obvious outdoor locations are the garage or the shed. The garage can be an integral part of the house or it can be 'stand alone'.

In the last two vicarages we have lived in the garage was connected to the house and that was extremely convenient. However, there are, as always, advantages and disadvantages.

The first obvious decision to make is what to do with the car. That may sound silly but these days many people never put the car in the garage because they use the garage for storage or as a workshop or simply can't be bothered to put the car in.

The location of the garage is also a major consideration. My elder son's, for example, is at the bottom of the garden and, to get to it with the car, he has to drive along a narrow alley way onto which the door opens. Many of his neighbours choose not to do this and park their cars on the road in front of their houses. Also, you have to get power to the garage if you are to use it for modelling and, whilst this is not an insuperable problem, any electrical work must be done properly and safely.

Another consideration is the type of garage you have. I would say cast concrete ones are not suitable for model railways, although some modellers have built layouts in such garages successfully. Damp on the internal walls can be a problem in winter and the walls are not suitable for fixing baseboards or supporting framework to them. However, David Jenkinson, who is a very well known modeller, built his "Little Long Drag" 4mm layout in such a garage **(Photographs 5/1 & 5/2),** which proves it can be done.

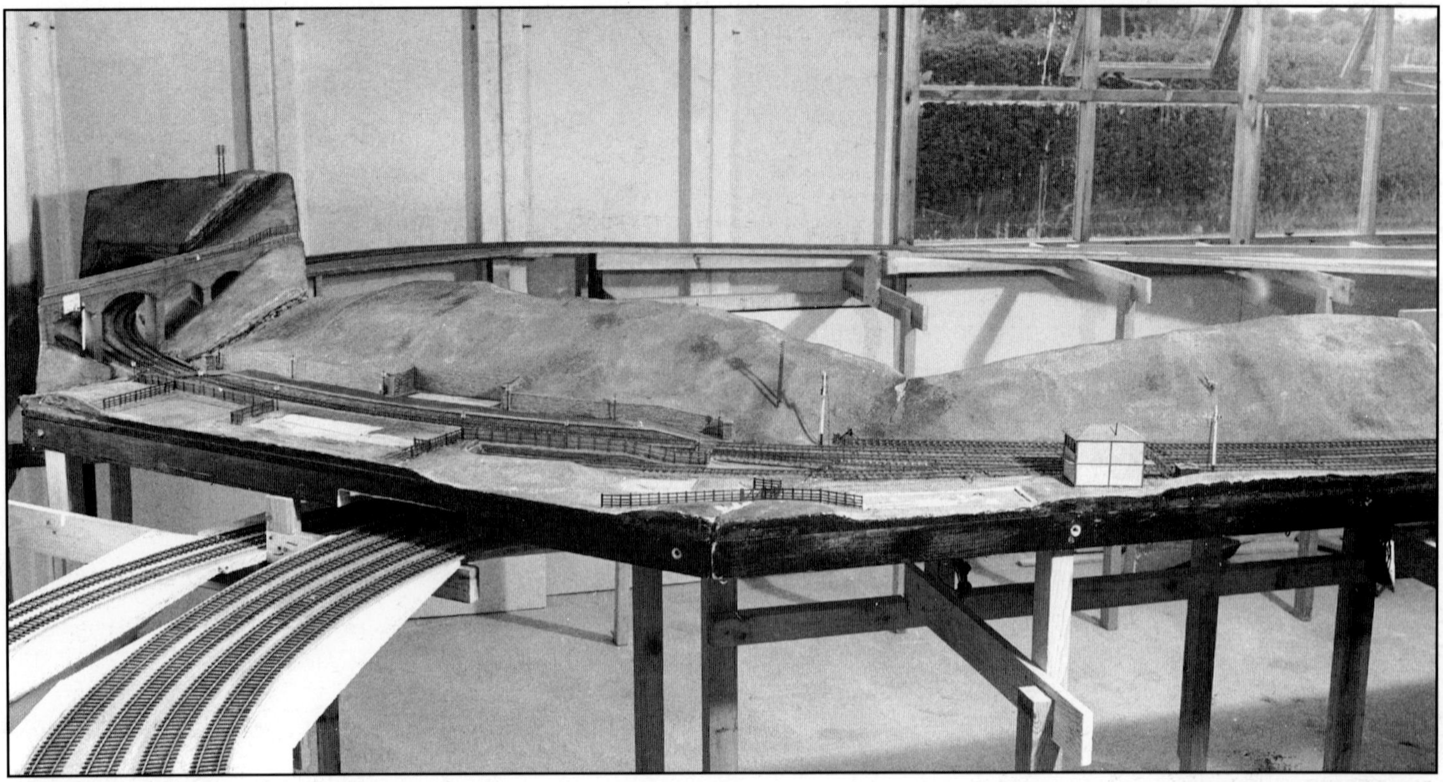

Photo. 5/1
A view of David Jenkinson's 'Little Drag' layout under construction with the cast concrete sections clearly shown in the background. ***Ron Prattley***

Photo. 5/2
Another view of David Jenkinson's 'Little Drag'. This time a train is passing through Dent Station. ***Ron Prattley***

CHAPTER FIVE

Wherever your garage is located, being usually of single skin construction, some form of heating will need to be install and draughts kept out from under the door(s). You may also need to add some insulation. If you choose to put the car in the garage you will have to design the layout in such a way that the car can be accommodated too. This obviously depends on the size of the garage and there have been many ingenious ways of keeping the car in the garage and housing a layout there as well. Finally, remember that a wet car increases the damp level in the garage and, if I had a layout in the garage with the car in it too, I would dry the car off before going indoors. In fact, I do this anyway because I think it is good for the car. However, I accept that, for many people, this would be an unacceptable extra chore particularly if you have had a long journey and it is late at night. All that being said, many people have very successfully housed their layout in the garage.

Fig. 5/1

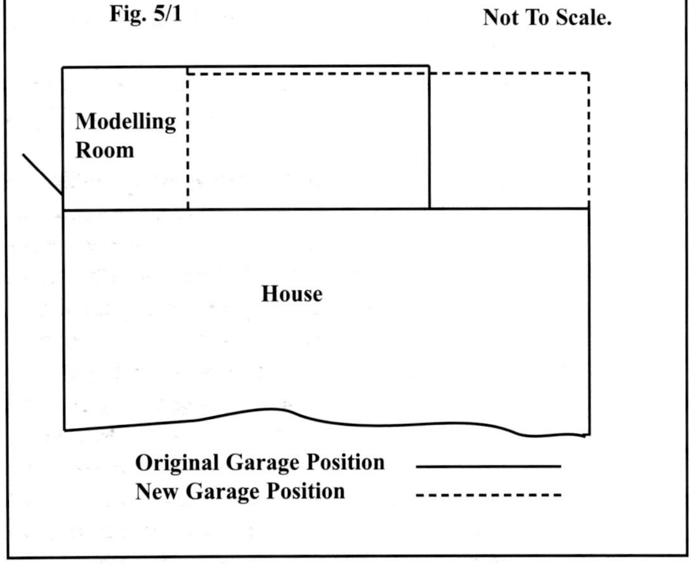

I built an 'O' gauge 16.5 layout myself in the rear of the garage in a house we lived in for 12 years in Leamington. I solved the problems by extending the garage slightly towards the front of the house and putting in a new breeze-block wall towards the rear to create the room. The garage was attached to the side of the house and slightly set back. All I did was get a builder to extend the garage forward so that the front of it was level with the house front and then build the new wall (See plans in **Figure 5/1**). The existing rear door to the garage became the entrance to the new room. To make the breeze-block wall look better and to add insulation, it was covered with hardboard with a surface that looked as if it was panelling. This was attached to battens and in the space between the wall and the panelling there was insulation. This proved to be a snug modelling room and, had we not had to move, I would have been very happy with it.

The plan of the layout is shown in **Figure 5/2** and a photograph with a much younger author **(Photograph 5/3)**. The floor was covered with underlay and carpet and heating was by an electric tubular heater for background heat augmented by a fan heater when necessary.

The other popular outdoor location for a layout is the garden shed and it is one that I have used on three occasions.

Again the shed has advantages and disadvantages.

The advantages are:

- The layout can have a permanent home.
- You can leave work and the layout in an unfinished state without having to tidy everything away.
- The layout and all that goes with it is self-contained.
- It provides storage space.
- It encourages modelling in good weather in summer months.

The disadvantages are:

- It can be cold in winter.
- Electricity has to be taken from the house or garage.
- It can be damp.
- It separates you from the family.
- Security can be a problem these days.

In common with many other modellers, I overcame most of the disadvantages but it cannot be denied that there is truth in the statement that it separates you from the family but that is true of the loft too.

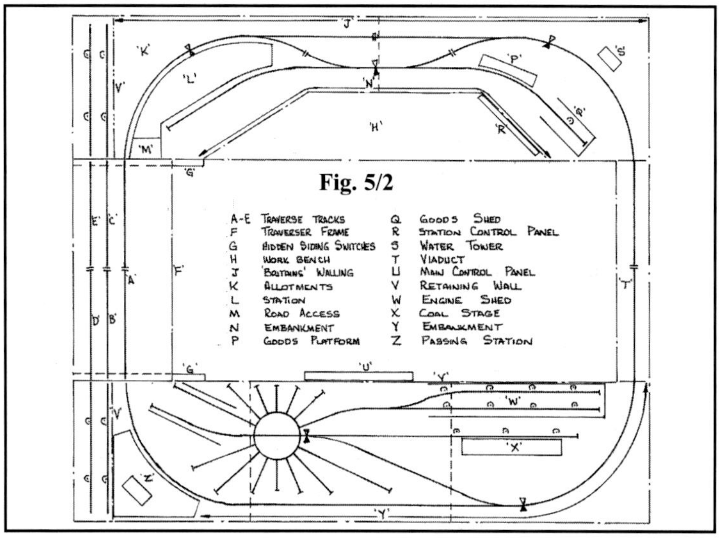

(Below) Photo. 5/3

"All Saints Depot". Note the wall panelling on the rear wall. The walls were painted white for added light. The kitchen cupboard on the wall was used for storage. ***The Late J. Prestidge***

Depending on the location of the shed, there is another possible disadvantage. If it is raining hard and the shed is a long way from the house, you will get wet getting to it!

You can overcome the problem of being separated from the family to a certain extent by doing jobs like kit building, timetable planning etc. indoors. On the other hand during the summer months when children are playing outside you can be in the shed working with the door and window open and readily available if needed.

I overcame the problems of a shed on each occasion that I have used one for modelling by:

Putting the shed on bearers. In one case they were well creosoted sleepers which had just been lifted nearby!

Regularly coating the shed with preservative.

Installing the electricity supply from the garage which was close to the shed. The cable was suspended on a stretcher bar and once inside the shed it went into a 'splitter box'. One branch supplied power for the layout and heating, the other for the lighting via an on/off switch. **IF IN DOUBT CONSULT A QUALIFIED ELECTRICIAN, BETTER TO BE SAFE THAN SORRY.**

Lining the shed interior. I pinned polythene sheeting to the shed sides between the battens and then using special adhesive I stuck 1" insulation to the polythene sheeting. This was a great help in summer and winter and, being white, the insulation made the interior much lighter.

Carpeting the floor.

Heating with a tubular heater

Putting curtains (made by my wife) to the windows.

It goes without saying that the shed must be on a level base. I have used either a full concrete base on which to rest the bearers on one occasion and twice I have used 2ft x 2ft paving stones which covered an area larger than the shed so that there was a patio surrounding it.

As far as security is concerned, you can only do your best and consult the local crime prevention officer if you want specialist advice. I have to admit that the problem has probably worsened over the 35 years or so since I first used a shed for railway modelling but, that said, many people do use a shed for their hobby.

Some people use the shed as the "covered terminus" for a layout round the garden and this can be very effective. In fact if you have a large enough garden you can have another shed or sheds for further stations, whether through or termini.

Many modellers who carry out their modelling in a shed would agree that using a spare room in the house is quite expensive in terms of space for domestic use and it would not be the first time that a layout in the spare room has had to be dismantled because of an unexpected addition to the family, or an elderly relative having to come and live with their son or daughter.

A shed, of whatever size, can be a reasonable investment for your modelling activities spread over a number of years. Sheds also have the advantage that they come in a variety of sizes with doors and windows positioned to suit you. Therefore, if you have an idea of the layout you want to build, you can go out and look for the shed that suits you best. The space underneath the layout can also be used to advantage for storage. On two occasions I have used discarded kitchen units as the means of supporting the layout and they provide excellent storage. They need not be yours either. The first time I used them they were being thrown out by a neighbour who was having a new kitchen fitted and was only too glad to get rid of them.

The same applies of course if you are building a layout in the garage or the spare room. Never let kitchen units go if you can find a home for them!

If you do not want to use kitchen units on which to support your layout or you cannot obtain any, the timber shed has the advantage that you can fix your baseboard frames to the sides and the floor very easily. Equally, storage shelves above the layout are easily fixed too. If you are thinking about having more than one level to your layout, don't forget that you can fix a narrow board for a terminal station or fiddle yard to the shed sides.

In my opinion, a shed, properly insulated, lighted and heated has much to commend it and if it is a choice between a layout in a shed and no layout at all, I would choose the shed. As an indication of how comfortable a shed can be to work in, the last time I used a shed for my modelling was when I was a curate. We lived in a typical 3 bedroomed pre-war semi-detached house and there was no spare room. When I moved to be a Vicar, my successor in the house found the shed so snug and comfortable that he made it his study.

Someone I know bought a shed and located it IN THE GARAGE. This made it much easier to install electricity and to get to, because the garage had direct access from the house and, as it was inside the garage, it was that much warmer. Of course the disadvantage was that there was no natural light but that is true of many lofts too. **Photograph 5/4** shows the very first garden shed I used from 1966 to 1970. It was 6ft square and cost £40.00!

Photographs 5/5 & 5/6 show 2 views of the late Dennis Doyle's layout "Ynysbyygwr" which was described in the Railway Modeller some years ago. If I remember rightly it was properly insulated and lined inside with suitable floor covering and was very comfortable with suitable heating. The garden shed behind the bus and lorry towards the top right hand side of **Photograph 5/6** is a model of the shed in which the layout was built.

(Left) Photo. 5/4
My first shed for the railway in 1966. ***Author***

Photo. 5/5
A view of the late Dennis Doyle's layout showing what can be achieved in a garden shed.
Ron Prattley

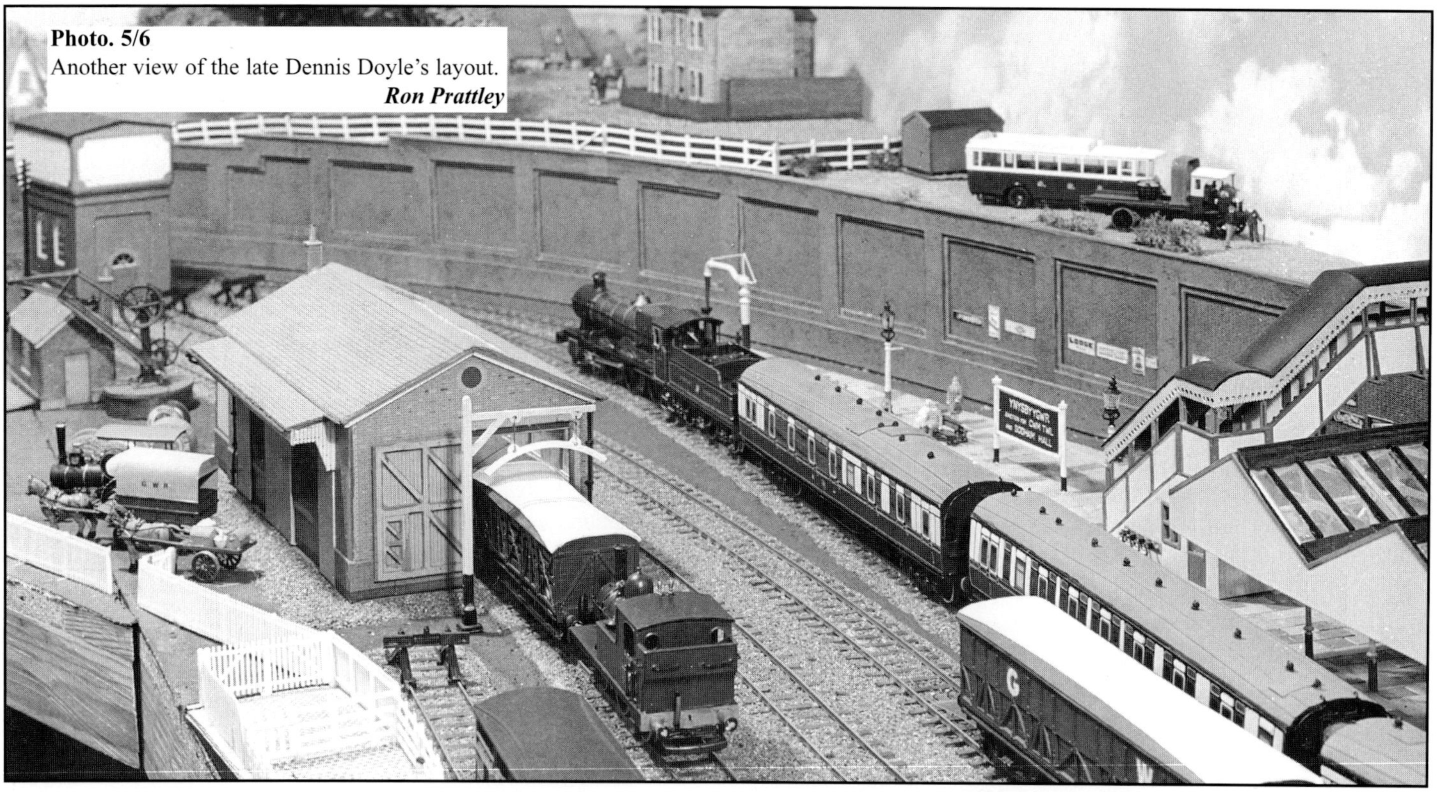

Photo. 5/6
Another view of the late Dennis Doyle's layout.
Ron Prattley

Another outside location is a conservatory, although as it is normally attached to the house, there is usually direct access whatever the weather conditions. The obvious disadvantage is the potential extreme temperature in the summer because of all the glass. However, this has not stopped people successfully building layouts in conservatories and, whilst I have no personal experience, they must have found ways of overcoming it.

There is no doubt that a conservatory is likely to add value to your house in the event of your selling it.

The other possibility for a self-contained layout in the garden is to look for a good second hand caravan. Depending on the space available, this could be a tourer as towed by many people or a mobile home. Some years ago in August 1986, there was an article in the Railway Modeller about using a touring caravan as a permanent home for a layout which went into some detail.

The unique advantage of such a location is its ease of moving your layout in the event of your having to move house.

I have never enquired about the prices of second hand caravans but I know there are dealers within easy reach of most towns and villages. Therefore, if you are attracted by the idea, you can make enquiries and look around potential caravans before finally deciding.

Another possibility is to use an outside store. These can be found in houses of all ages. Many pre-war semi-detached houses had two, usually joined together, at the rear of the house. Often they were made into an extension by knocking through from the kitchen, but equally they could be used for railway modelling. The room created by knocking the toilet and the coal store into one would be long and thin but that space is not an obstacle. You just have to design a layout to suit the space available. A typical fiddle-yard to terminus layout springs to mind and there would be room for storage underneath and possibly room for a small bench. (see **Figure 5/3**)

The proximity to the house and the fact that they are joined to it makes the installation of electricity that much simpler.

In the last vicarage in which we lived there was a coal store outside the kitchen as an integral part of the house and a garden store at the rear, also an integral part of the house. The latter was at the rear of the house next to the garage and in both cases it would have been possible to build a layout in one of the smaller scales in them. Again, being an integral part of the house, it would have been easy to install electricity. I can't remember the exact dimensions but, from memory, the garden store measured about 6ft x 4ft and people have built layouts in that space in the past and will doubtless do so in the future. (See **Figures 5/4 & 5/5)**

If you have money to spare and want to add value to your property, then consider building an extension. I have done this once and it provided good extra usable space. We used it for play space for our two sons by providing a playroom each. That was necessary because one son wanted to build a layout whereas the other was into "military modelling". A luxury it may be but, when we moved, the people that bought the house were attracted to it because it provided them with "his" and "hers" hobby rooms! **(Figure 5/6)**

Fig. 5/3

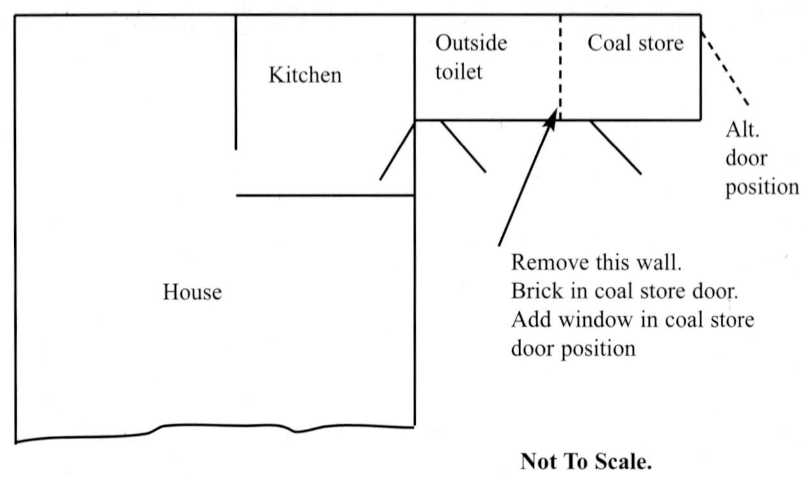

Fig. 5/4

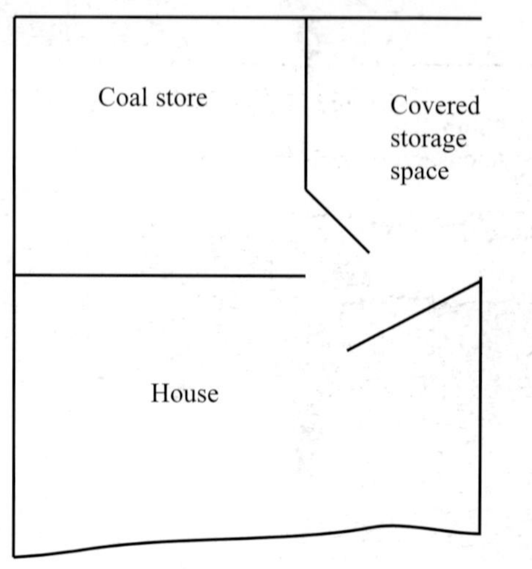

Fig. 5/5

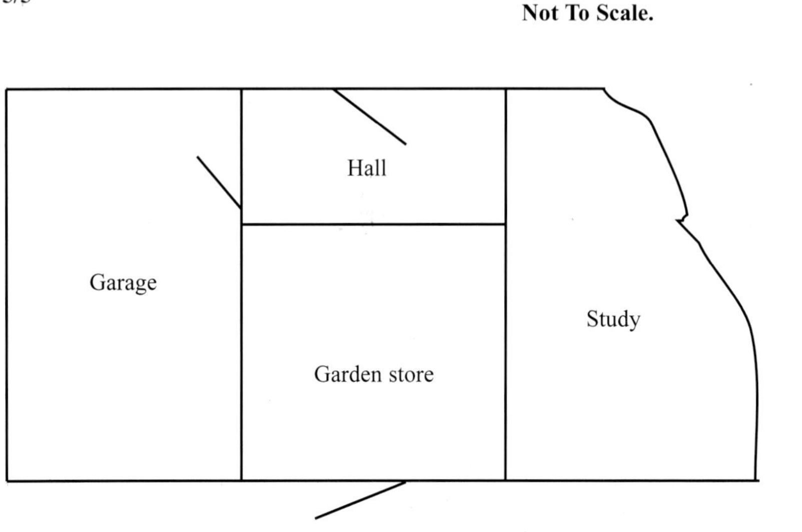

Fig. 5/6

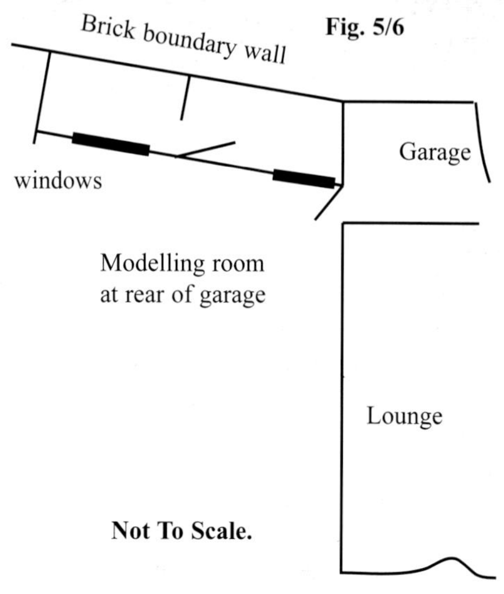

CHAPTER FIVE

John Allison, a prolific layout builder and member of the Warley M.R.C. has an extension to his house and has located at least two layouts there over the years which have been described in the model railway press. "Seven Mile House" and "Oval Ash C" (See **Figures 5/7 & 5/8** for plans and **Photographs 5/7 & 5/8**)

Finally, you can have covered outdoor layouts if the garden of your property allows it. The easiest way to describe this is by the block plans in **Figures 5/9 to 5/14**. The cover can be provided by fencing with a Perspex roof over, for example, or by whatever means you choose. The layouts would go in the shaded areas. It has to be said that you need a shed in which to store rolling stock etc. or somewhere in the house or garage. However, many people with garden layouts do this and I think it is an option worth exploring if you want a layout and cannot have one elsewhere.

(I have deliberately not discussed permanent layouts whether indoors or outdoors. This is because it is self-evident that, if you have found the space for such a layout already, you will not be concerned with answering the question "Where can I put my layout?" which is the subject of this book.)

Key to the plan of Seven Mile House.

C = Capstans (optional)

H = Stone crusher

L = Loader or tippler

R = Repair shops or factory

T = Vertical traverser (could be a sector plate)

S = Storage roads (fiddle yard) fed from traverser or sector plate.

O = Embankment or retaining wall

P = Plate girder skew bridge

X = Critical headshunt

E = Possible to extend this headshunt if space allowed

Q = Control & operating position.

Fig. 5/7 Track Plan of Seven Mile House

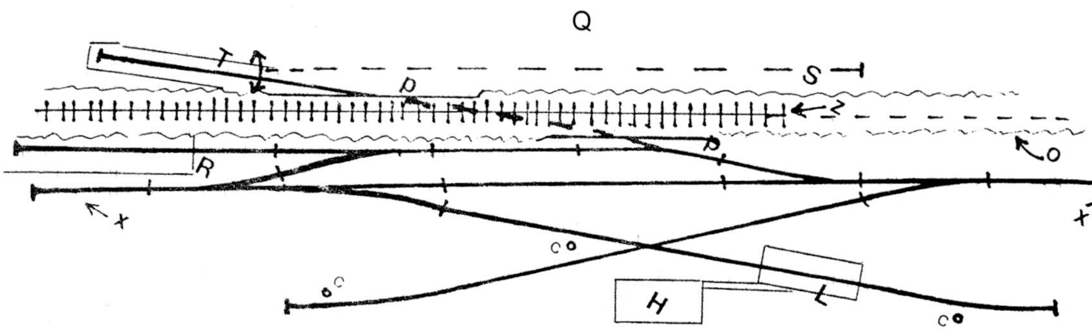

(Above) Fig. 5/7
This layout was built in 'O' gauge but it is possible to build it in 'OO' gauge in a space of say 6ft x 2ft. If you have a bigger area avaible use the same plan but have larger sidings. In 'N' gauge you could build this layout on an ironing board!

(Left) Photo. 5/7
If you have the space, you could make the baseboard rectangular and extend the layout where it curves round towards the TV set in the photograph.
John Allison

Fig. 5/8

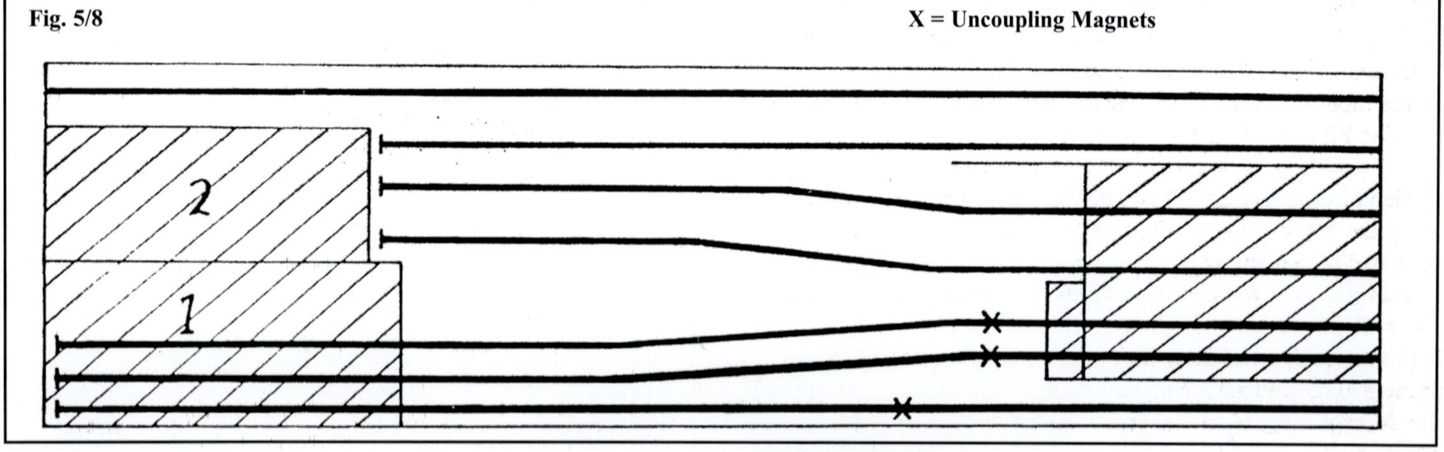

Photo. 5/8
Note the layout (Oval Ash C) is on top of the piano. It is also in two parts so that it can be exhibited. The hinge at the front has a removable pin & there is a similar one at the back. You can see how the power is carried across the join via the plug and socket.

John Allison

Fig. 5/9

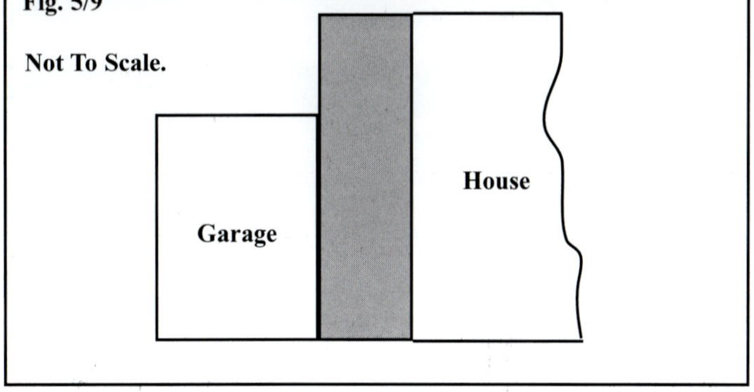

Fig. 5/10

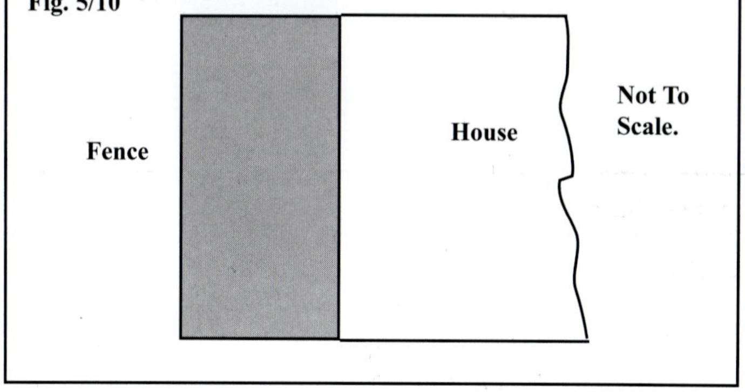

Fig. 5/11

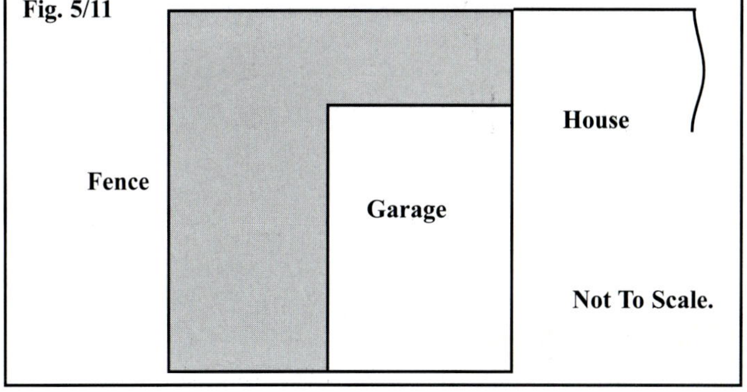

Fig. 5/12

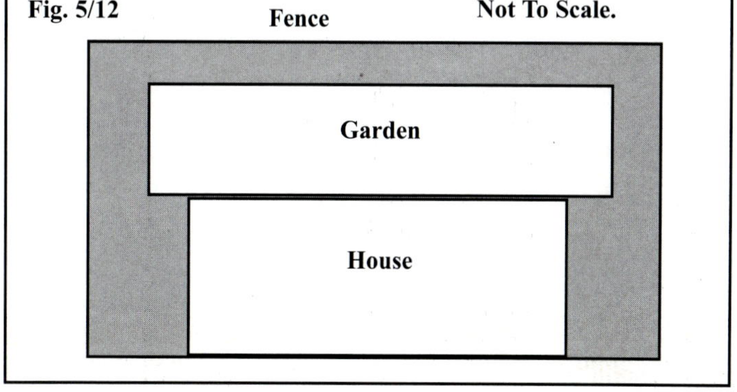

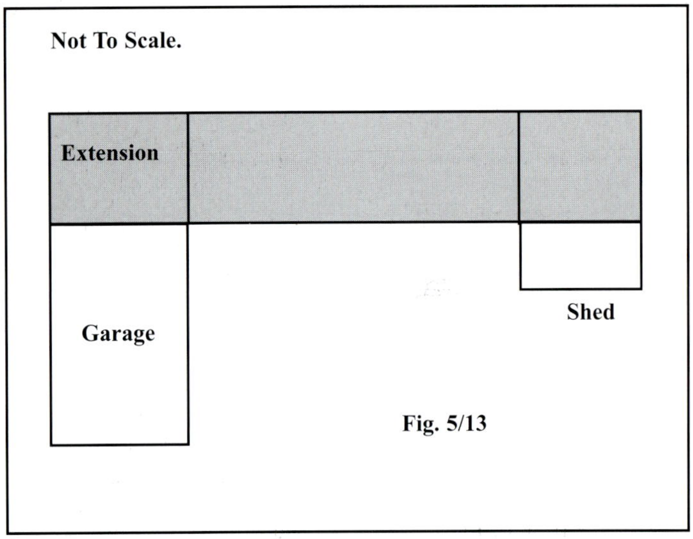

Fig. 5/13

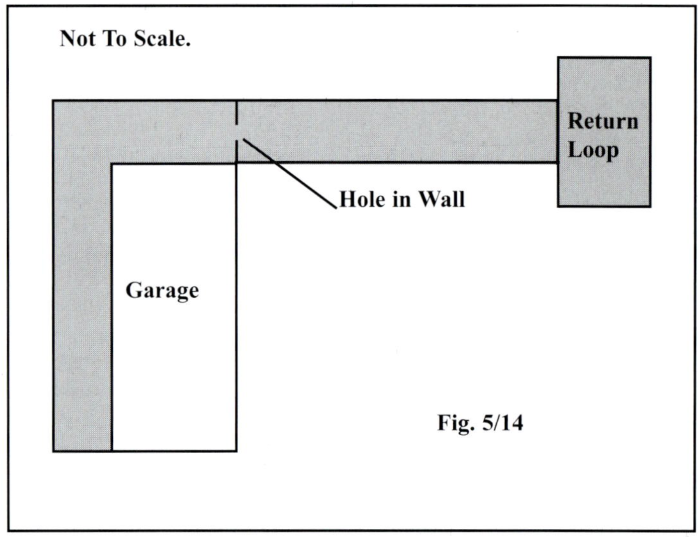

Fig. 5/14

6. Workshop Space.

If you are going to build a layout you will need some space to perhaps make track, build rolling stock, paint figures, make buildings and so on. But you do not need a workshop in which to do this. If you can, that is marvellous but it is certainly not essential.

Many first class models have literally been made on the kitchen table. For example, some years ago, our local Focus store was selling workbenches which were designed to fold away when not in use. They were very simple in construction and just needed to be fixed to a vertical surface (shed or garage). I have not seen them lately so perhaps they are no longer made but the principle holds good.

The typical workbench you find in a shed or a garage is not good for modelling in my opinion because you need to be able to sit down to work. I would find it most awkward, not to say tiring, to have to stand at the bench to do modelling work.

An article by Rod Neep in MRJ No.5 (published in 1986) describes how he made himself a portable workbench, including storage space, from white melamine faced chipboard. It was very effective and he took it to exhibitions too and used it for his demonstration stand. I also know of a modeller who bought a cheap bureau and used that as his workbench and the drawers for storage. This has the advantage that you can simply shut the lid when it is time to stop modelling, sure in the knowledge that everything should be where you left it when you go back to it. If you can only find a table to work on but nowhere else to store your tools, then a table top fold-up workstation is for you. A number of these have been described in the model railway press over the years.

A very good one is described in the American book mentioned earlier – "Model Railroading in Small Spaces". My own bench is an old computer workstation which we bought when our sons were much younger and we were going to discard it until I realised its potential. I added two trays for tools and covered the main flat surface with formica. I also put two lengths of track on the top narrow surface and here I also have some storage containers. A 4-way socket is permanently fixed to the left hand side and a vice to the right hand side and, lastly, the formica is covered with 2 cutting mats. Overall the workstation measures 33 inches wide x 21inches deep. It might be thought to be small but it suits me admirably and has done for a number of years. See **Photographs 6/1 and 6/2.**

Photo. 6/1
A view of the workbench. ***Author***

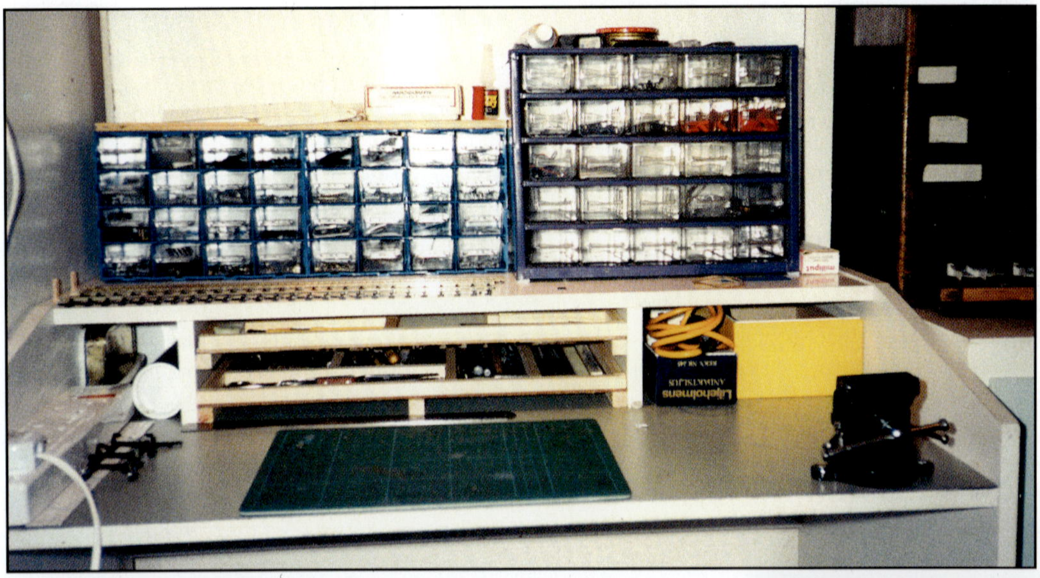

(Left) Photo. 6/2
A further view of the workbench. ***Author***

(Below Middle) Photo. 7/1
The purchased storage box with the lid closed. ***Author***

(Below Bottom) Photo. 7/2
Note the raised strip in the middle of each section. This is the same as the 'back to back' measurement between the wheels. ***Author***

7. Stock Storage.

Clearly, if space is at a premium, it can be a problem to store your rolling stock because most of us have too much stock for the size of the layout we have. I know that is true in my case.

In my opinion, nothing is worse than keeping stock either in their original boxes or simply wrapped in "bubble-wrap" and kept in plastic or cardboard boxes. It is not economical in space terms and the risk of damage is much greater especially if you intend to exhibit your layout. I have to say I never cease to be amazed at the way some people carry their stock to and from exhibitions and still expect it to work, let alone the risk of damage.

There is only one way to store stock effectively and that is in proper stock boxes. These can be purpose built and I have one such box **(see Photographs 7/1 & 7/2)**.I bought this one but the principle for all stock boxes is the same. Have dividers the correct distance apart to allow you to fit the stock between them. Foam rubber is an added luxury or necessity depending on your viewpoint. However, do not run away with the idea that you must either buy or purpose build stock boxes.

Mine have come from a variety of sources:

A picnic box **(Photograph 7/3)**

A wooden box with a fitted lid made many years ago from plywood by my father-in-law to house a record player.

Two cardboard boxes I bought at an exhibition, sold, if I remember rightly, for storing 'O' gauge coaches.

Two toolboxes with 3 purpose built trays in each made from MDF board **(Photographs 7/4 & 7/5)**

Two small metal filing cabinets (the sort you carry around) with 2 purpose built trays in each, again made from MDF board. **(Photographs 7/6 & 7/7)**

An old small case **(Photograph 7/8)**

In the case of the small filing cabinets I made out a cutting list, took it to Focus and they cut all the bits of MDF for me at a nominal cost. I took them home and glued and pinned them together.

The photographs speak better than any diagrams or further words. As an additional luxury, to ease transport to and from exhibitions I bought two folding plastic trolleys on wheels with extending 'towing handles' as fitted to suitcases. In one I put the stock boxes I am taking and in the other I put three plastic boxes stacked one on top of the other which carry the remaining necessities one takes such as transformers, controllers, lighting bulbs, tool box etc. The trolleys only cost £9.99 each (in January 2003) from Focus and I think they are first class because when not in use they simply fold up.

CHAPTER SEVEN

As I get older anything that makes it easier to carry stock boxes and other things in and out of exhibition halls has got to be a good thing! These plastic trolleys could also be used for storage under layouts. **(Photograph 7/9)** Alternatively they can be used only for taking stock to shows and stored folded when not in use. They can, of course, also be used for other household transport jobs.

Having suggested various homes for your layout what follows in the next chapter is a selection of layouts which either I or other people have built.

One thing certain about the hobby of railway modelling is that it has been in existence for so many years that it is doubtful if anything new can be invented!

We all, consciously or sub-consciously, copy other people's ideas and, in a sense, that is why magazines like Railway Modeller and British Railway Modelling are so popular. Modellers describe their layouts or their projects and when we read them we are so impressed that we either copy or modify the idea to suit our own circumstances.

Of course, there have been original ideas put forward over the years and there may well be in the future but, as far as I am aware, no modeller who writes about his or her work has ever put a patent on anything! They tell us about their models in the hope that we find it useful. I know that this is the case with what I have written in magazines in the past. If someone can benefit and use any idea of mine I shall only be too pleased because I will have put something back into the hobby which I have enjoyed since I was a child. If your layout has to be small, that's fine. I particularly like building small layouts anyway! Some people might think that if a layout is small the interest will quickly wane because it doesn't take so long to build, but I and many others find that adding scenic detail is very absorbing. I have included some photographs both of the prototype and of layouts which show this. These are shown in Chapter 9 which is simply a collection of photographs and captions.

(Top Left) Photo. 7/3 Picnic Box.

(Top Middle) Photo. 7/4 Toolbox with tray.

(Top Right) Photo. 7/5 Toolbox tray.

(Middle Left) Photo. 7/7 Inserts for filing cabinets.

(Below) Photo. 7/9 Plastic folding trolley.

(Above) Photo. 7/6 Small filing cabinet.

(Right) Photo. 7/8 Old small case.

8. Some Actual Layouts.
The Yard

Roger Lincoln has built a 4ft x 1ft American HO scale switching layout called "The Yard". For those not familiar with American terminology 'switching' is shunting. It was originally intended as a design exercise but was built and completed and is now exhibited.

It can be carried by one person and can be stored almost anywhere. Roger has stored it in a wardrobe or under a cabin bed. **Photograph 8/1** shows it stored in the wardrobe. The layout has two carrying handles fitted to the underside. The legs on the right hand side are, in fact, for another layout but could be used with "The Yard" if required.

Alternatively, if the layout had folding legs, it would be self-contained and the space at the right hand side could be used to store another layout. The top shelf is used to store the controller, extension lead and enough stock to operate the layout. This is better seen in **Photograph 8/2**.

Roger has also utilised the storage area under a cabin bed to good effect. This bed is mounted 4ft 6ins. above the ground and Roger currently stores under it 3 layouts, 3 unused 4ft x 2ft 6ins.

(Below) Photo. 8/3
Storage under the cabin bed
Roger Lincoln

(Above) Photo. 8/1
"The Yard" stored in the wardrobe. ***Roger Lincoln***
(Below) Photo. 8/2
Storage of stock, controller on wardrobe shelf. Note one of the carrying handles on the layout ***Roger Lincoln***

baseboards, 6 stacker boxes, 2 chests of drawers on castors which are pulled out for access and various other bits of modelling equipment.

The 4 shelves at the end of the bed are also used for storage of model railway equipment and some Scalextric **(See Photograph 8/3)**. Note too that the double socket has been moved from the wall to the bed unit so that cables do not trail across the floor. In the same room is an alcove fitted with shelves for further storage. **(Photograph 8/4)**. The alcove is nicely fitted out with 12mm MDF shelves strengthened by the triangular supports. The stored items are in clearly labelled boxes and tidy. The latter point is particularly important if the room is used for purposes other than modelling, although, in my opinion, tidiness anywhere can only be a good thing. If you wished to hide the shelves it would be easy to fit a curtain.

(Left) Photo. 8/4
Alcove Storage.
Roger Lincoln

(Below) Photo. 8/5
A bird's eye view of **"The Yard"**.
Roger Lincoln

As far as "The Yard" is concerned the plan can be seen in **Photograph 8/5** or in **Figure 8/1**. It is standing on the 'Workmate' it was built on and on which it is operated at home. **Photograph 8/6** shows that much use has been made of British kits – Wills, Ratio and Airfix – as well as American bits.

Much attention has been paid to the track. The sleepers have been weathered and the track painted rust colour and the ballast is poor as is often the case in yards compared with the main line. The track at the back will be used for a further extension. **Photograph 8/7** shows that, by taking the scenery into the fiddle-yard area, the join between the two is less noticeable. It also gives more operating potential and makes the layout look larger. The buildings in this photograph are inverted Heljan goods shed, a much hacked about DPM kit and some scratchbuilt to suit the sites. The convincing weathering of the track helps disguise the fact that it is Peco Setrack.

Photograph 8/8 is a wide angle shot which makes this small layout look huge! The layout uses the Kadee coupling system and the magnets are visible because Roger couldn't face the hassle of recessing them! It has the advantage of being able to easily see where they are. Not everyone would go along with this and would prefer them to be recessed. In that case, their position would have to be marked by an item of scenery, e.g. a figure.

Fig. 8/1 Plan of "The Yard"

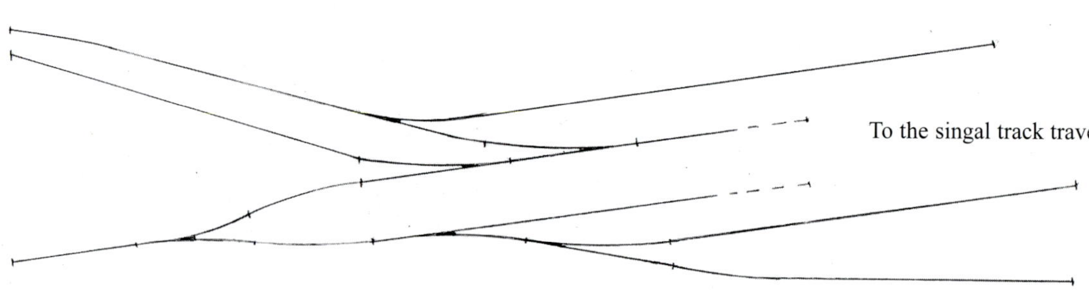

Photo. 8/6
A view of "The Yard" showing the use of British and American bits used in the construction of the layout. Notice also the detail on the two platforms. The tank on the roof looks particularly effective. The black oblongs in the two nearest tracks are the uncoupling magnets ***Roger Lincoln***

Photo. 8/7
This view is interesting because it shows the scenery has been taken into the fiddle yard area and makes the "join" less noticable. The postioning of the uncoupling magnets needs to be decided carefully to optimise the operating aspect of the layout.
Roger Lincoln

Photo. 8/8
The low relief buildings look very good and show clearly how easy it is to create a realistic backscene when the baseboard is narrow and there is not much space available for scenery.
Roger Lincoln

CHAPTER EIGHT

Layouts by Revd Wilbert Awdry.

About 30 years ago when 009 layouts seemed to be much more popular than they are now, probably due to the fact that rolling stock was readily and cheaply available, Rev'd Wilbert Awdry (author of Thomas the Tank Engine books) built a layout in that scale called "Ulfstead Road". This was what was called a 'rabbit warren' layout. Such layouts were so called because the dived in and out of tunnels on a number of levels in a small space.

"Ulfstead Road" was part of the fictional Mid-Sodor Railway and the track plan is shown in **Figure 8/2.** This, together with all the stock fitted in the boot of an Austin 1300 to travel to the many exhibitions at which it was shown. **Photographs 8/9 & 8/10** give an idea of what the layout looked like. It is still possible to obtain 009 track and rolling stock so, if this sort of layout appeals to you, why not build one? The layout was operated to a tape recorded commentary.

Fig. 8/2 Track plan of Ulfstead Road.

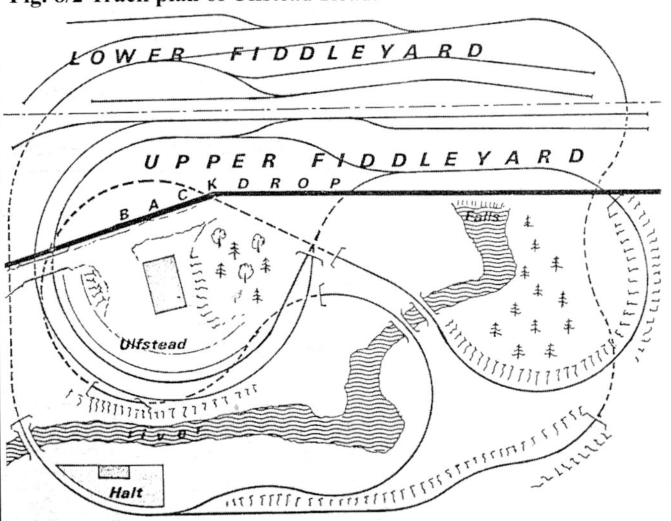

(Below) Photo. 8/9
The model trees of today are probably more realistic but in its day the trees on this layout were good. ***Christopher Awdry collection***

As I said in Chapter 4, about 50 years ago it was commonly thought that you should build a layout for the average home on a solid baseboard measuring 6ft x 4ft or 8ft x 4ft. Many such layouts were built but they had major disadvantages, not least their weight and the fact that they were very difficult to move about easily.

Wilbert Awdry was one of the few people to build such a layout which was successfully shown at many exhibitions. He called the layout "The Ffarquhar Branch" and the plan is shown in **Figure 8/3. Photographs 8/12 & 8/13** show how effective a layout it was especially remembering that the majority, if not all, of the buildings used were made from proprietary kits. I remember this layout and its builder with great affection as Wilbert brought it to the very first exhibition which I organised in December 1966. It was also the Oxford M.R.C's first show.

Nearly 13 years later he was exhibiting "Ulfstead Road" at the M.R.C. Exhibition in Central Hall, London, when Justin (my elder son) and I were exhibiting a layout we had built to show what father and son could do with proprietary equipment. It was called "Brookhurst" after Justin's school. Justin was 9 years old and, not unnaturally, was rather overawed at operating a layout in front of so many people. However, a few kind words about the high standard of his operating from Wilbert made him feel 10 feet tall and he never looked back.

(Above) Photo. 8/10
"Albert" and train entering Ulfstead Road.
Christopher Awdry collection

In later years, Wilbert built "Ffarquhar Mark 2" which was an 'end to end' layout with two fiddle yards, one of which was hidden behind the station. **(See Figure 8/4). Photograph 8/11** shows an overall view of the layout taken at Rodborough Model Railway Exhibition in September 1978 and the two fiddle yards can be clearly seen.

The layout was hinged in the middle and the two halves folded together for transport. In **Photograph 8/15** you can see the high level signal box and the water tower which were removable. They cover the hinges and the planked walkway between them disguised the join between the two baseboards. Not surprisingly, Wilbert featured the locomotives and rolling stock from famous books on his layouts and **Photograph 8/14** shows Thomas approaching Ffarquhar station with Annie and Clarabel. Thomas was a rebuilt Hornby E2 and the coaches were Exley LNER suburbans.

I am sure there is a 'market' for a good exhibition layout based on Wilbert's (and now Christopher's) books. My sons enjoyed the books when they were small and now my grandchildren do too. When I say 'good' I mean a good scenic layout properly operated as distinct from a 'tailchaser' where the trains travel at a scale of 100mph.

Photo. 8/14

Photo. 8/15

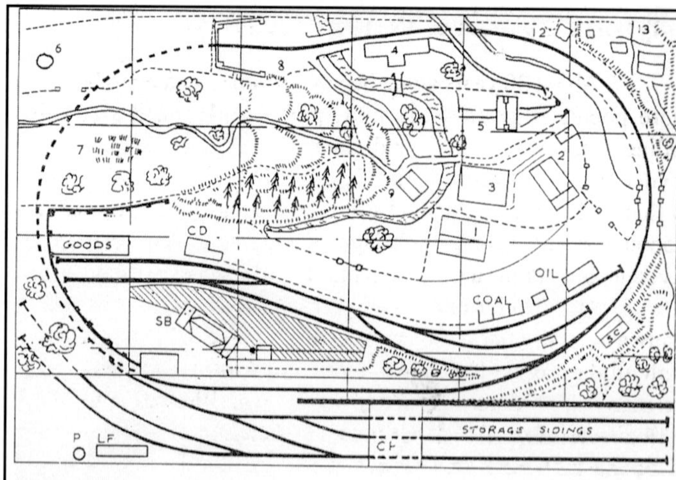

(Above) Fig. 8/3 Plan of the original "Ffarquhar Branch"

(Left) Photo. 8/11
A general view of Ffarquhar Branch Mk2 taken at Rodborough Exhibition in Sepetember 1978 with Wilbert at the controls.
Christopher Awdry collection

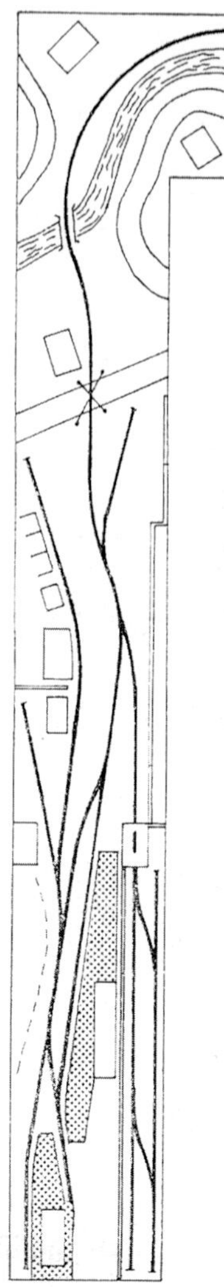
Fig. 8/4 Ffarquhar Mark 2

Photo. 8/12 ***Christopher Awdry collection***

Photo. 8/13 ***Christopher Awdry collection***

(Far Left) Photo's 8/14 & 8/15
Notice the detail on a layout built over 30 years ago. Of particular note is the high level signal box disguising the entrance to the hidden storage sidings. It is the top part of the Airfix (Dapol) kit. I have just done something similar on my latest layout.
Christopher Awdry collection

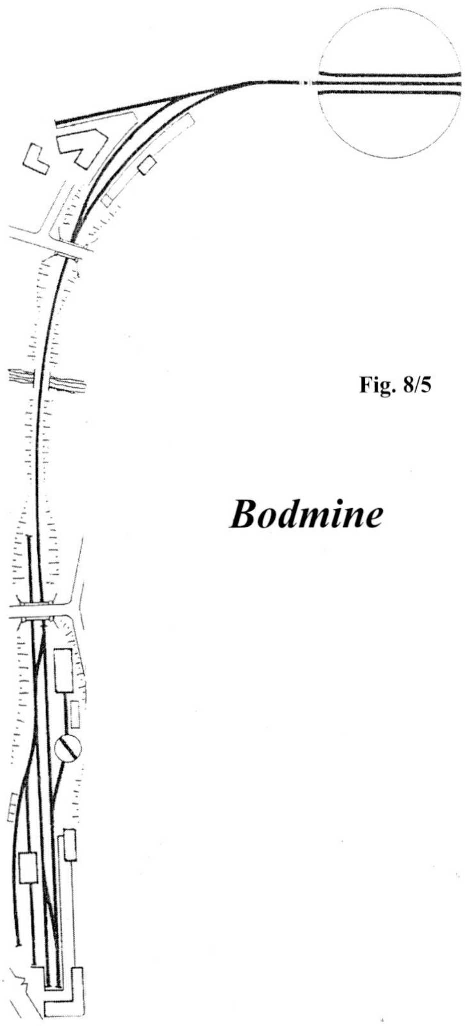

Fig. 8/5

Bodmine

Bodmine.

Another layout, but this time in 2mm scale, is John Greenwood's "Bodmine". This started out as a board measuring 4ft 9ins. x 1ft simply because John had a piece of chipboard that size and it ran onto a plank of wood as a fiddle yard. He quickly found this was inadequate and built two extra baseboards which took the track onto a rotating fiddle yard **(See Figure 8/5)**. The boards are joined together by nuts and bolts and alignment is by pattern-maker's dowels.

The rotating fiddle yard is a turntable deck of 22ins. diameter on which is laid 5 tracks. Such fiddle yards can be operated by hand and located by eye and then held in place by a bolt. John's is operated by a 12 volt motor driving a rubber tyre against the lower surface of the deck via Meccano gears to turn it. Alignment with the entry track is via a pivoting cam with a step and a switch on it. This latches onto stops on the underside of the deck to stop and locate it.

The track is handbuilt and the points are mechanically operated by rods and cranks underneath the baseboard connected to a lever frame. The electrical side of the layout is simple due to the simple nature of the track plan. To start with, John used kits and bits to get as much as possible done quickly but he gradually replaced most of these. This raises an important point. Just because a layout is complete, you don't necessarily have to sell it or break it up and start another one. You can improve what you already have if you prefer. John's initial plan was to build a small scenic test track but he ended up with a complete small layout which was exhibited regularly. Originally it was housed in the loft but John moved and it now lives in a large cellar. Before the move he had plans to extend the layout round the loft and build another station so that he could have an 'end to end' layout.

Photographs 8/16 8/17 & 8/18 show how realistic the layout is.

(Below) Photo. 8/16
A goods trains passing through the countryside on Bodmine.
Tony Wright

(Above) Photo. 8/17
The buildings, the old lorry and the general scenery give a superb impression of the English countryside. ***Tony Wright***

(Left) Photo. 8/18
Another view of the house shown in the above photo. This time it shows the front aspect. Notice one of the windows is open and there are even curtains. All this in 2mm scale. ***Tony Wright***

WENFORDBRIDGE GOODS DEPOT.

Whilst Bodmine was still being exhibited at shows, John started building his model of Wenfordbridge Goods Depot. Experience had shown that Bodmine was very heavy to lift and so the baseboards for the new layout were made of plywood and open framed to keep the weight down. Again the track was hand built and many of the scenic techniques and ideas John used on this layout are described in detail in Barry Norman's book "Landscape Modelling".

The main feature of the layout is the working gantry crane. Containers are transferred from arriving Conflat wagons to a Scammell mechanical horse & trailer which then goes off down the road to disappear through a hole in the backscene. All this is done without any hand appearing from the sky and is very impressive. The track plan is shown in **Figure 8/6** and **Photographs 8/19 & 8/20** show what a superb layout this is.

John has now joined Wenfordbridge to Bodmine in his cellar.

Fig. 8/6 Wenfordbridge Goods Depot.

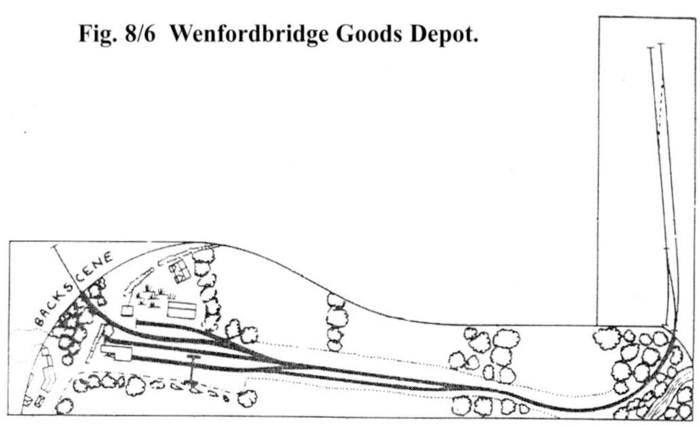

Photo. 8/19
Many people would think this "for real", not a 2mm scale model. ***John Greenwood***

Photo. 8/20
A close up of the working crane unloading the log onto the trailer. ***John Greenwood***

Kenilworth Junction.

It is very common for modellers to "move up a scale" as they get older. Perhaps it is to do with having more space because the family have grown up. Perhaps it is to do with our eyes not being able to cope with small scale models as we get older or it might be because we just prefer the larger scale. In my own case the answer probably lies in a mixture of all those reasons.

Another modeller who has followed this road is Alan Searle. In the 1970's he was modelling in 3mm scale and built a superb model of Kenilworth Junction. In real life the junction was formed by the divergence north of the double track line to Berkswell on the Coventry - Birmingham line as shown in **Figure 8/7**. Alan lived less than 300 yards from the junction for the first 26 years of his life so he was well qualified to build a model of it. The track plan is shown in **Figures 8/8** & **8/9** shows how the layout looked when it was being exhibited. This latter drawing is important because it shows how you can display drawings and photographs of the area modelled so that visitors to the exhibition can relate the model to the original.

The baseboards were made of substantial 2in x 2in timber covered in 1/2in chipboard and would have been very heavy. Alan would not use that today! The track was GEM flexitrack and the points were electrically operated by the 'electric pencil' method. The crossover was, of necessity, handbuilt in situ.

I include this layout because it shows that you can build a very convincing model of a prototype in an area which is not too large to be accommodated in an average house.

3mm scale modelling may not be as popular as it was 30 years ago but it has its faithful followers and they build some very good layouts and models. If space is at a premium, 3mm scale is well worth considering.

Photographs 8/22- 8/24 show how effective the layout was and **Photograph 8/21** shows the prototype in about 1960.

(Left) Photo. 8/21
Kenilworth Junction c.1960. All that exists now is the bridge and the single track to Coventry. *Alan Searle*

(Below) Fig. 8/7
Track plan of Kenilworth Junction.

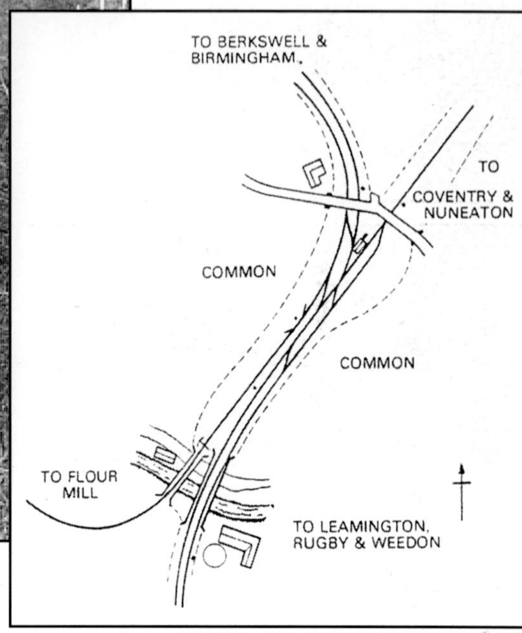

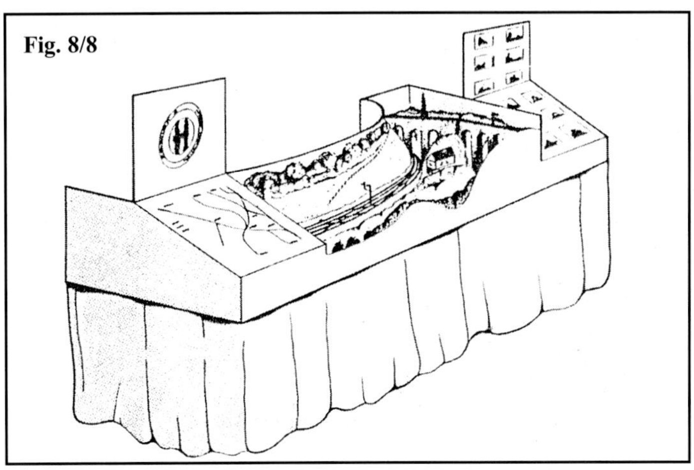

Fig. 8/8

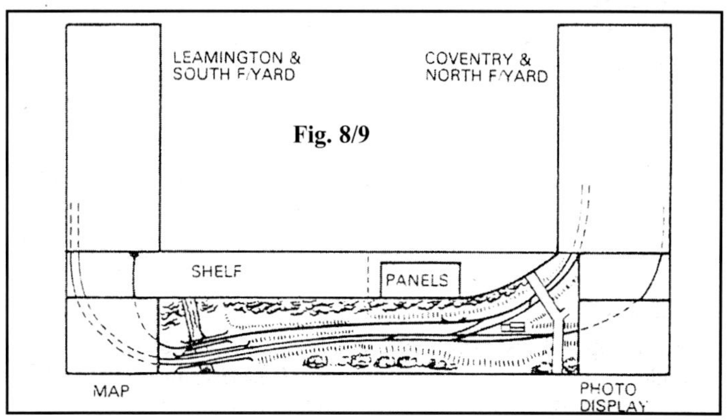

Fig. 8/9

Photo. 8/22
The 'Jinty' waits for the parcels train to pass. The scenery is very effective. Notice how the fence follows the contours of the track, just as it did in real life. ***Alan Searle***

Photo. 8/23
Kenilworth Junction signal box. A superb model of the protoype. If you have the skill to do this it is very satisfying and, if you don't try, you will never know. ***Alan Searle***

Photo. 8/24
Stanier 'Black Five' picks up the token for the signal line to Coventry. Compare the model of the bridge with the prototype in **Photo 8/21**. It is an excellent reproduction of the real thing. Notice too the allotments on the right hand side with the bean sticks. ***Alan Searle***

Stoke Fleming.

Alan's current 'O' gauge layout is a minimum space one measuring 12ft x 2ft. The baseboards and integral legs are made almost entirely of scrap plywood and can be erected or dismantled in approximately 5 minutes.

The plan is shown in **Figure 8/10** and the photographs, **(8/25 to 8/29**), show how effective the layout is. It is not surprisingly very popular on the exhibition circuit and has featured in the Railway Modeller.

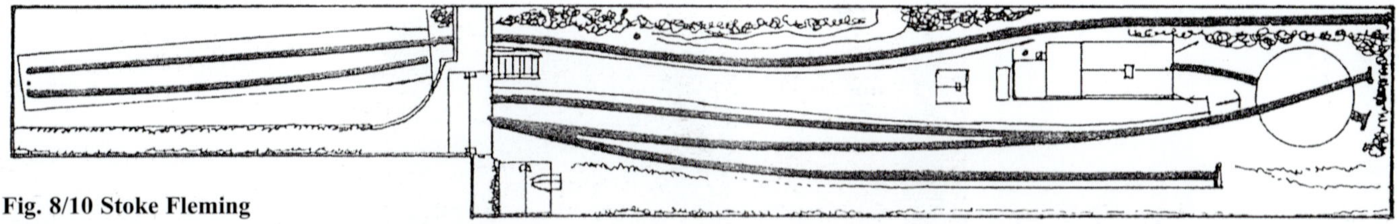

Fig. 8/10 Stoke Fleming

Photo. 8/25
The evening sun casts shadows over the turntable and watering facilities at Stoke Fleming. Just look at the scenic detail in this photograph. I would think most people would believe this was the 'real thing' rather than a model. ***Alan Searle***

Photo. 8/26
The 'Prairie' is cast in the evening sun as it is turned on the turntable at Stoke Fleming. Notice the damaged fence wire so commonly seen but very rarely modelled. The change in the type of fencing is also a good point to note as it adds variety. ***Alan Searle***

Photo. 8/27
The scratch built engine shed is very effective. Notice the fire buckets hanging just inside the door. The downpipe is not quite straight which adds to the authenticity.
Alan Searle

Photo. 8/28
The signal box is mounted on the platform reminding us that this was quite often done. From a modelling point of view, it saves precious space. Notice too the flower beds at the base of the two station nameboard supports.
Alan Searle

Photo. 8/29
An overall view of the end of Stoke Fleming platform. The loco is running around the brake van, a passenger stands on the platform, the signal box window is open and the station cat enjoys the sun outside the box. A typical country station picture of days long gone. *Alan Searle*

Whitcombe & Pensevern.

If you want to build a layout in a shed and exhibit it, you can either make the whole layout portable or do as Ray Hazell did over 30 years ago and make the majority fixed and part removable to take to shows.

In this case, Whitcombe was 'fixed' and Pensevern, which was the branch line, was transportable. It measured just over 11ft long and was 12inches wide at the station end and 18 inches wide at the other end. When being exhibited it had its own separate fiddle yard which was stored in the shed. Of course, when in the shed, Pensevern was joined to Whitcombe. The shed measured 12ft x 8ft and was supported on old railway sleepers. It was lined and insulated with white expanded polystyrene sheet. This was also attached to the roof which, in addition to the insulation properties gave excellent reflective assistance to the lighting. The baseboards were, as one would expect for late 1960's, made of 2in x 1in timber topped with insulation board. These were joined together by bolts and the whole layout rested on a frame work built on and braced to the sides of the shed. There was plenty of room for storage under the framework.

The track was hand built from half track kits and the points were electrically operated from a central control panel with panel mounted controllers and all the switches for feeds and sections.

The town and station buildings were mainly from kits modified where necessary.

There were lots of figures on the layout – over 120 on Pensevern section alone – which adds life to the scene.

The plan of the layout is shown in **Figure 8/11**. It was an attempt to create a layout based on GWR West of England practice. Pensevern still exists stored in Ray's loft.

Photographs 8/30 - 8/33 show various views of Whitcombe. In particular, **Photograph 8/33** shows how scenery was mainly from Superquick & Bilteeze Kits.

Ray no longer has photographs of Pensevern but it was described in the December 1970 issue of the Model Railway Constructor, which you may be able to pick up if you are interested.

Fig. 8/11
Track Plan Of "Whitcombe & Pensevern"

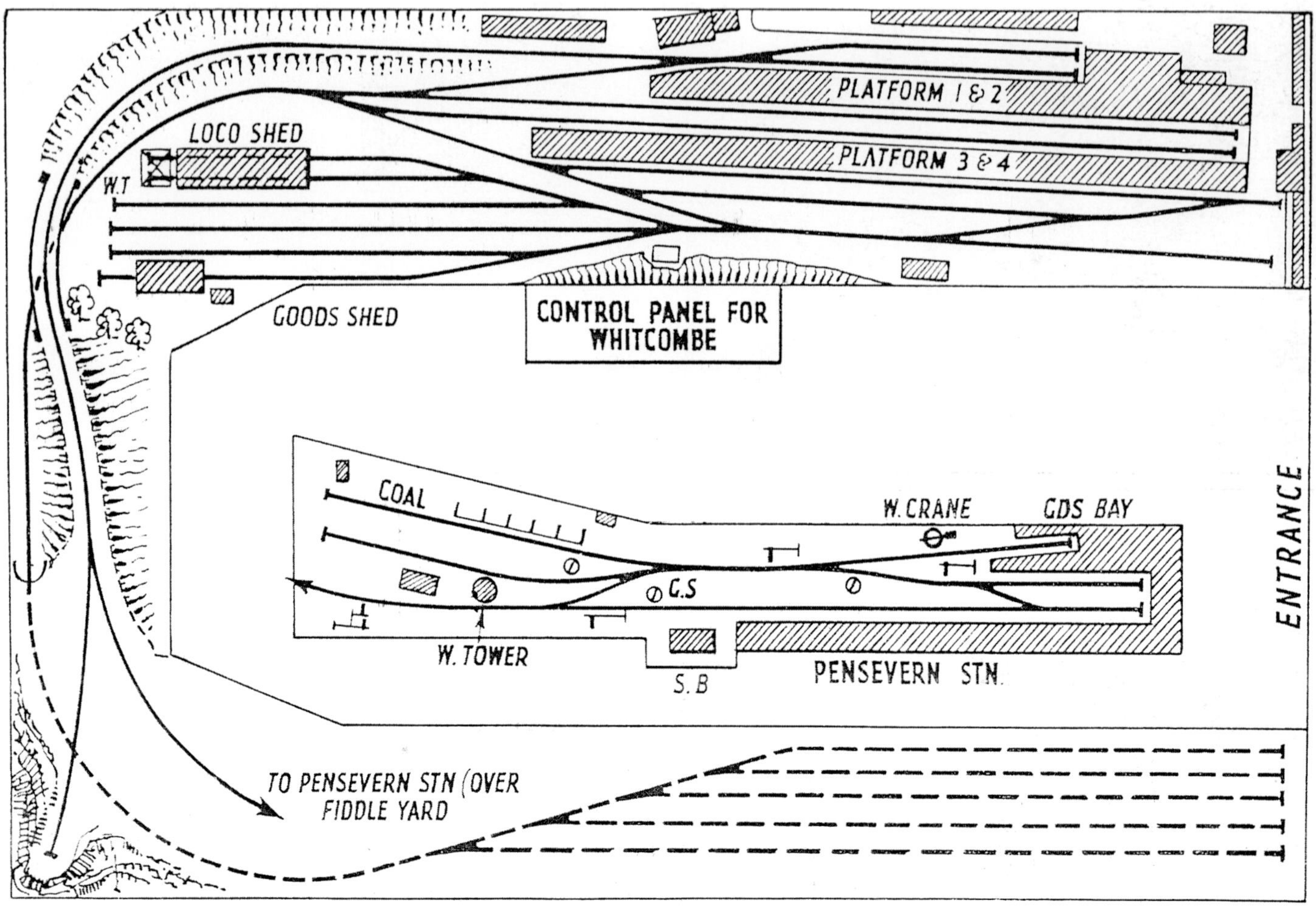

Photo. 8/30
A view across Whitcombe, with a Castle ready to depart and an Auto-train standing in the bay having arrived from Pensevern. *Ray Hazell*

Photo. 8/31
A overall view across Whitcombe, with a pair of Castles standing at the platforms with their trains and the Auto-train arriving from Pensevern. *Ray Hazell*

Photo. 8/32
The line behind the signal box is the branch to Pensevern. The viaduct is just visible behind the engine shed. *Ray Hazell*

Photo. 8/33
A view across Whitcombe Station and the streets around it. *Ray Hazell*

Abingdon

More recently Ray has built a scale model of the GWR terminus at Abingdon. This branch ran from Radley, on the Oxford to Didcot line, to Abingdon providing a passenger and goods service.. Originally the line was broad gauge and it was converted to standard gauge in November 1872. Abingdon was also the home of M.G. Car Co. and cars were transported by rail to various parts of the UK at one stage.

The track layout of Ray's model follows the original plans taken from the British Rail/Oxford Publishing plan.

All the points are hand built by Ray and operated by H & M motors. The station building is constructed from plasticard and built to plans in OPC "Great Western Architecture" by Adrian Vaughan. The goods shed is hand built using photographs and information gathered when Ray visited the site, then practically demolished, in the 1980's. Foundations were still available to photograph and the track and platforms were still in situ.

The layout is 16ft long and 3ft wide at the station end tapering to 2ft wide at the loco shed end.

Information for the scenic background was obtained from photographs taken by Ray in the 1980's and from books and magazines specifically:
1 Railway Magazine July 1982 article by N.E.Trippett entitled "A Trumpery Little Railway".
2 "Abingdon in Camera" by M.J. Thomas
3 "The Abingdon Branch" by N.E.Trippett & Nicholas de Courtais published by Wild Swan Publications Ltd.

This model is very special to Ray because he was born in Abingdon and his mother's family lived there. His mother's brother, Frank Moor, was one of the drivers on the branch line which was affectionately known as the "Abingdon Bunk".

There were two crews who alternated "earlies" and "lates".

Photographs 8/34 - 8/37 show various views of the layout.

Photo. 8/34
A overall view across Ray Hazell's Abingdon *Ray Hazell*

Photo. 8/35
A close up of the main station buildings and platform at Abingdon. *Ray Hazell*

Photo. 8/36
A view across the goods yard at Abingdon. *Ray Hazell*

Photo. 8/37
A view of Abingdon's cattle dock. A Dean Goods shunts in the yard behind. *Ray Hazell*

Extended M.P.D.

In Chapter 4, I briefly mentioned that I had built a small M.P.D. in 'O' gauge and then extended it when we moved to Nuneaton. The baseboards were of conventional design and, as can be seen from the photographs, there were no points on the layout. This has advantages as all you have to do is wire feeds and dead sections and of course you save considerably in cost because of the lack of points.

In a loco yard the tracks are not ballasted in the normal sense. I reflect this by mixing Polyfilla with black powder paint and 'trowelling on the resultant mess' with an old knife and smoothing it out with my forefinger. Any which sticks to the rails is easily removed with an old screwdriver. Personally I think it looks effective and, where the rail sides are visible, they are painted rust colour. I have done this on a number of layouts where I have modelled a loco shed.

The major feature on such a layout must obviously be the engine shed. This one had two. The original board had a half relief one made from the large Engine Shed kit produced by Heljan (and since released again). The second board has a shed made from two of the subsequent Heljan kits which were of thicker material and smaller in size. I used about 1 1/2 kits and the unused parts were used as low relief buildings for some of the backscene.

The track that ran behind the shed was covered by an overall roof (made from the H.O. Goods Depot by Lima) and a water tank (Airfix) which were removable for transit. This track was in effect another fiddle yard because locos stored there could not be seen until they appeared from behind the shed.

On a layout like this I believe it is essential to concentrate on scenic detail. Some of it, like the benches inside the shed, was not very visible. Some was very small like the two newspapers being read, the birds on the roof and the black cat but, for me, that added to the fun. "Fun" is important and so I added Del-Boy's van as I am a fan of "Only Fools & Horses".

If you are exhibiting a layout, one sure way of attracting people's attention to the scenic detail is to challenge them to find certain items. This is easily done by putting up a list which says :-
"Can you find ....?"

People will spend ages doing this.

The list need not include all the detail on the layout. Examples on my list include:

Birds on the roof and wall
Gas bottles
Man & stepladder
Man reading paper
Man working in the drain

I have written in the past tense about this layout because it is no more. I broke it up because I needed the space but I was able to salvage everything except the track and boards. So the switches and the scenic items are currently being recycled.

Photographs 4/45 and **4/46** on page 28 together show an overall view of the layout. It consisted simply of 4 parallel tracks on each board joined by a traverser. It was literally a 'pointless layout'.

Photo. 8/38 An overall angled view of some of the detail on the new board. The use of containers for storage is quite common in many yards. Note too the various figures, including the man in the orange "safety overalls", and the bench, stepladders, etc.

Photo. 8/39
Concentrates on the graffiti on the fence (by my younger son Paul) but also note the detail on the railway side of the fence. This was on the original small M.P.D. board before the layout was extended.
Martin Hewitt

Photo. 8/40
Another overall view of the extension to the original layout showing the shed and, in front of it, a lot of scenic detail.
Martin Hewitt

Photo. 8/41
A close view showing Del-Boy's van and the skip by the wall. You can just see the black cat sat on the grey generator trailer in the corner. The bike leaning against the skip adds a typical touch.
Martin Hewitt

Photo. 8/42
A close up of the man working in the drain with 2 other men. One is the boss, talking to him. The drain cover is to the right.
Martin Hewitt

CHAPTER NINE

Prototype Scenic Details.

(Above) Photo. 9/2 Minffordd Station on the Ffestiniog Railway. I include this to show the different colours of ballast and sleepers, the fence, apparently being re-painted because it is two different colours, the flaking paint on the water tower & the tied up water hose. ***Author***

(Above Top) Photo. 9/1 A traverser for trams at Crich Tramway Museum. The same setting could be used for a wagon or loco works. Note the cobblestones/ granite setts. ***Author***

(Above) Photo. 9/3 The yard at the I.O.W. Steam Railway Works. There is a wealth of detail here including a rotary washing line! ***Author***

(Right) Photo. 9/4 Bridge supports at Ryde St. Johns (I.O.W.). Also note drainpipes painted different colours. ***Author***

(Left) Photo. 9/5 The loco shed at Ryde St. Johns. Note the roller shutter doors and numbers over them, the bricked up round window on the building to the right, the shop sign and the different ground levels. ***Author***

(Left) Photo. 9/6 Llanfair Station on the Welshpool & Llanfair Railway. Note the corrugated iron building. ***Author***

(Below) Photo. 9/7 The coal stack at Llanfair, surrounded by a "fence" made of sleepers. Notice all the blue plastic "buckets" loaded with coal ready for the next loco to be re-coaled. ***Author***

(Below) Photo. 9/8 The loco shed at Pendre. A good use for old wagons and a short siding. ***Author***

(Below) Photo. 9/9 The Talyllyn Railway. Behind the shed at Pendre. There is an amazing amount of detail to be seen here. Rotting sleepers, an old lever frame, wheelsets, bits of metal etc. ***Author***

(Below left) Photo. 9/10 Oil drums at Pendre on breeze blocks outside the North Carriage Shed. Note the other "clutter", and building work must be going on, see the pile of sand and the shovel. ***Author***

Model Scenic Detail.

(Left) Photo. 9/11 Another example of layout detail provided by suppliers. All I did was to put the various bits together in a cameo. Notice the man on the left with a mug of tea in his hand and his newspaper on the bench.
Martin Hewitt

(Below) Photo. 9/12 This idea came from an American magazine. I made the sign using my computer and the small signs are purchased in sheets, cut out and stuck to plasticard.
Martin Hewitt

(Left) Photo. 9/13 Men at work watched by the boss looking out of the window. All figures etc. are easily bought and make all the difference. *Martin Hewitt*

(Above) Photo. 9/14 A conversation taking place. Again figures, cylinders and drums are easily obtainable.
Martin Hewitt

(Left) Photo. 9/15 The engine shed, the diesel refuelling point and all the clutter add up to a realistic scene.
Martin Hewitt

Postscript.

Finally, what about those people who really do not have space for a layout? My suggestion is join your local club. If there isn't one, form one! I have done that 3 times now. I was the founder member of the Oxford MRC in 1966, the Leamington & Warwick MRS in 1974 and the Tywyn MRC in 2001. I was also a founder member of 'COVGOG' in 1992. It is not difficult. A notice in the Railway Modeller, British Railway Modelling and other magazines and a short piece in the local paper will, I am sure, produce enough interest to form a club. In my experience finding somewhere to meet has not been a problem but I accept I have been lucky. If you don't try you will never know.

I would also encourage readers to build a layout which can be exhibited. I have been exhibiting layouts for 36 years and it really is enjoyable. If it's not your own, help to exhibit your club layout. You meet a lot of fellow modellers, see a lot of layouts and get plenty of ideas for your own layout. In my opinion, exhibitions are essential for the continuation of our hobby. Every club exhibition I have organised has always produced some new members and you never know how many seeds you have sowed amongst the young (and not so young!). Also belonging to a club and visiting shows means that you can pick people's brains.

Happy modelling, wherever it is and wherever you are.

Photo. PS/1 The Author at an exhibition, in Solihull in 1979, concentrating on operating his '009' MPD. You can just see one of the old fashioned type "dolly switches" used on the control panel.
Author's collection

Photo. PS/2 Three members of the Tywyn M.R.C. working on the club layout (Aberdysynni) at the 2003 exhibition where it was shown under construction. ***David Mitchell***

Photo. PS/3 Another view of the Tywyn M.R.C. 2003 exhibition. The layout in the foreground is Alan Searle's "Stoke Fleming" and in the background Justin Adams & Earl King are operating Earl's layout.
David Mitchell

Photo. PS/4
A photograph of some members of Oxford M.R.C. taken in the early 1970's in the clubroom with the club layout which was a model of Radley on the Oxford-Didcot line. Sadly John Pomroy (on the extreme right at the back) has since died. The author is third from the left behind the layout. At the time he was modelling in OO9 but still helped to build and operate Radley. This is one of the advantages of belonging to a club.
Author's collection

Photo. PS/5
In many ways this photograph and the previous one could be said to be what the hobby of railway modelling is all about. The photo of the Oxford MRC members shows people who meet together, share the same hobby and build a layout that individually they would never have built.This photograph was taken at the 1979 Leamington MRS exhibition. This layout is 'Brookhurst' built by me and my elder son, Justin , who is on the left of the picture. You can see from his shirt that the layout had been shown at that year's Model Railway Club Exhibition at Central Hall Westminster. A much younger author stands at the back, Justin's friend Ian is next to him and the two other boys were members of the public. By working together and exhibiting our layouts we encourage people to take up an absorbing hobby. Personally, I always hope we attract younger people to ensure the hobby continues for many years to come. ***Author's collection***